# Macro-engineering: MIT Brunel Lectures on Global Infrastructure

It is a massy wheel,
Fixed on the summit of the highest mount,
To whose huge spokes ten thousand lesser things
Are mortis'd and adjoin'd; which when it falls;
Each small annexment, petty consequence,
Attends the boist'rous ruin.
Shakespeare: *Hamlet, II, iii*

# Macro-engineering: MIT Brunel Lectures on Global Infrastructure

**Frank P. Davidson, Ernst G. Frankel and C. Lawrence Meador** (Editors)

Macro-Engineering Research Group
The School of Engineering
Massachusetts Institute of Technology
Cambridge, Mass
USA

**WP**

WOODHEAD
PUBLISHING

Oxford   Cambridge   Philadelphia   New Delhi

Published by Woodhead Publishing Limited,
80 High Street, Sawston, Cambridge CB22 3HJ, UK
www.woodheadpublishing.com

Woodhead Publishing, 1518 Walnut Street, Suite 1100, Philadelphia,
PA 19102-3406, USA

Woodhead Publishing India Private Limited, G-2, Vardaan House, 7/28 Ansari Road,
Daryaganj, New Delhi – 110002, India
www.woodheadpublishingindia.com

First published by Horwood Publishing Limited (formerly Albion Publishing), 1997
Reprinted by Woodhead Publishing Limited, 2011

British Library Cataloguing in Publication Data
A catalogue record for this book is available from the British Library

ISBN 978-1-898563-33-4

# Preface

In presenting this volume of the M.I.T. Brunel Lectures, the Macro-Engineering Research Group of Engineering acknowledges the considerable assistance and encouragement received from many individuals, laboratories, and departments of the Massachusetts Institute of Technology. The interdisciplinary study, preparation, and accomplishment of large-scale engineering works designed for improvement of the planetary environment and of the human condition could hardly be envisaged without teamwork on a grand scale.

For the opportunity to pursue this formidable task, we are grateful to President Charles M. Vest, to Provost Joël Moses, and to Dean of Engineering Robert A Brown. Beyond the metes and bounds of the Cambridge campus, inspiration and cooperation have been extended by the cadres of the International Association of Macro-Engineering Societies and especially by its founding helmsman --himself a Brunel lecturer represented in these pages -- Uwe Kitzinger, whose wide-angled lens has enabled him to serve, successively, as Economic Advisor when Lord Soames played such an eminent role in the European Community; as Dean of INSEAD, the European Business School established in Fontainebleau at the suggestion of the late General Georges F. Doriot, as President of Templeton College, Oxford and, more recently, as an "Explicator of Europe" at the Harvard Center for European Studies! *Ars celere artem.*

The editors wish also to acknowledge the expert and knowledgeable services rendered by Cherie Potts as coordinator, once again, of the preparation of the manuscript for publication. (She played a similar and equally indispensable role in the production of the first volume of Brunel Lectures, published in 1992, and in volumes of the Macro-Engineering Series brought out under the auspices of the American Association for the Advancement of Science.)

To our brilliant publisher, Ellis Horwood, MBE, we owe much in the way of direction and invaluable suggestions. He has been a wise and patient friend through all the vicissitudes encountered in bringing this multi-authored project to fruition. If macro-engineering is to take its rightful place as an "engine" of tomorrow's world, it will in no small measure be the result of interdisciplinary teamwork and élan. When difficulties arose -- and they did! --We were able to reply on the old Roman admonition: *de minimis non curat praetor.*

There is, to be sure, a long tradition of military construction of basic elements of what George Washington termed "internal improvements", now known -- in an age addicted to more complex terminology -- as "infrastructure". The Roman Army built the first extensive road system in Europe and North Africa. Thierry Gaudin, whose wise chapter concludes this volume, reminds us that the innovative use of iron for weapon-making in the early Middle Ages was soon followed by the fabrication of metal plows; this revolution in agricultural technology doubled the food production -- and the population -- of Europe.

In this *optique*, General Tom McInerney's initiative in adapting the software of the CTIS (Command Tactical Information System), used to such conspicuous effect in the Gulf War for the cleanup of the Exxon-Valdez oil spill, could be regarded as a traditional example of "evolutionary macro-engineering". In reality, the transition from a war-winning management system to a "peace-winning" nexus of computer and communication technologies required rare perceptiveness and persuasiveness as well as sound knowledge and persistent leadership. Gen. McInerney's key role was recognized by the presentation to him of the Isambard Kingdom Brunel Award. With characteristic modesty, Gen. McInerney carefully mentioned the individuals on his staff as well as others who had helped modify a brilliant piece of military software for civilian and international use. It all constituted vital assistance, at a very critical moment, to environmental defense and restoration. This significant effort was recently paralleled by an expansion of the telemedicine network of the United States Air Force.

Another former Brunel lecturer, Professor Manabu Nakagawa provided leadership and liaison in Asia for the cooperative understanding needed to extend the reach of the world's leading medical authorities -- now in a relatively small number of teaching hospitals and research laboratories -- to remote areas where local doctors can be informed and equipped to join the proliferating and humanitarian network of telemedicine.

As macro-engineering cases and principles become better known, one can hope for wiser selection of projects and programs, and for more rational and compassionate decision processes. André Bénard's moving account of the travails of arranging finance for the English Channel Tunnel might have had a more encouraging ambience if construction had been approved as soon as the first serious offers were received by the Channel Tunnel Study Group -- in 1959! -- to build this "shaggy dog" of macro-engineering projects! In 1959, the estimated cost was $100 million. Successive delays brought the ultimate cost to more than $15 billion. If only political and other leaders could have taken more seriously

> If it were done, when 'tis done, then 'twere well
> It were done quickly                          *Macbeth*, Shakespeare

One notices, particularly in the West, a self-defeating welter of administrative bodies which inevitably reduce competitivity, environmental health, and general amenity and good feeling. A case in point is the U.S. eastern seaboard's well-documented need for deep-water ports which could be constructed on artificial islands, as Nigel Chattey and his talented friends have instructed us. Why is it that Japan has been able to build more than 80 artificial islands -- admittedly their buildable space is more limited than ours -- while the United States has virtually none at all? Is it not time to establish a National Commission on Macro-Engineering, long advocated by John Landis? His reminiscent account of the Mulberry Project in the present volume (Chapter 2) could in some ways be regarded as presaging the floating, re-locatable islands described by Ernst Frankel in Chapter 8.

In the "tightly coupled" world announced years ago by Jay W. Forrester, it has become, however, a matter of urgency to improve and broaden our deployment of large-scale engineering. Science, wisely applied, should make it possible to halt the advance of deserts, to train young people more effectively for viable roles in a pan-technical economy, and to launch and sustain enterprises -- public, private, and mixed -- which increase wealth and amenity.

On the agenda for the near-term future must be a substantial effort to persuade the legal profession, at the highest level, to undertake -- perhaps as a restatement of facets of administrative law -- a simplification of the "approval process" for large-scale engineering. When as many as ten or twenty jurisdictions, each with its own timing and requirements, must separately pronounce on the merits of a novel system, the result is a predilection for delay. As leaders of the engineering profession prepared for the International Conference, "Macro-Engineering in the 21st Century", held at M.I.T. from October 24 through 27, 1996, there was renewed interest in the "delay factor", the long interval "'twixt cup and lip", while so many obvious tasks remain in the limbo of non-decision. Is Robert Goddard's design for supersonic trains to remain neglected, more than a decade after its essential validity was demonstrated on the M.I.T. Athletic Field? Must the African Sahel await the pumps and pipes for which the late Sir Robert Jackson pleaded at the Rensselaerville Institute a generation ago? Perhaps the time has come to name inter-sectoral task forces, so that the full power of that much-misunderstood nexus we call "the market" can be marshalled alongside the authorized agencies of government, to bring about both the infrastructural and the institutional changes so necessary if we are to reap the benefits adumbrated in the present volume.

Cambridge, Massachusetts
September, 1997

Frank P. Davidson
Ernst G. Frankel

# Contents

# EDITORIAL NOTE

The previous (first) volume of MIT Brunel Lectures covered the decade 1983-1992. Entitled *Macro-Engineering: Global Infrastructure Solutions*, it appeared in 1992 under the *imprimatur* of ELLIS HORWOOD LIMITED, a division of Simon & Schuster International Group. The present (second) volume of these special guest lectures, delivered before interdisciplinary audiences at the Massachusetts Institute of Technology during the five-year period 1992-1997, comes out in the Horwood Engineering Science Series of HORWOOD PUBLISHING LIMITED, the same publisher's new company.

A word of more than perfunctory appreciation is due, individually, to Mr Ellis Horwood, MBE, FRSC. His untiring and discerning attention to all aspects of the book provided the editors with very valuable guidance and – not least – the very broadening experience of close collaboration with a true professional publisher.

The passage of time takes its inevitable toll. We deeply regret the death of Monsieur Joseph Elkouby, author of Chapter 5. Long a principal advisor to the French Minister of Transport, his judgement and experience explain much of the "rayonnement" of the TGV network in Europe. Dr John Landis, retired from his key position in Stone & Webster Engineering Corporation, continues to contribute valuable services to the engineering societies of which he is an officer and director. Monsieur André Dénard, author of Chapter 3, retired from the chairmanship of EUROTUNNEL after seeing the project through to completion. Dr Andrew C. Lemer, author of Chapter 4, followed a brilliant term as Head of the Urban Research Board of the US National Research Council (Washington DC), by re-entering private life as CEO of the MATRIX Group Inc. In Baltimore, Maryland. Lieutenant-General Tom McInerney, after retiring from active duty with the United States Air Force, has been rendering distinguished services as a Senior Adviser to the Vice-President of the United States.

The editors, too, have "shifted gears". While continuing to carry a full teaching load at MIT, Professor Frankel has become Chairman of the Board of the Quincy Shipyard in Boston, Massachusetts. C. Lawrence Meador has devoted increasing time to his responsibilities as an officer of a major insurance company in Philadelphia, Pennsylvania. Frank P. Davidson has retired from MIT and serves as an officer and director of professional associations in the field of macro-engineering. *Ars celere artem.*

# 1

# The Suez Canal Re-Visited: 19th Century Global Infrastructure

**Jean-Paul Calon**
*Former General Counsel, Suez Canal Company*

The building of the Suez Canal is an excellent example of the ambition of our time, a macro-project to which everyone associated devoted the best of their thinking. This project was also a global infrastructure solution, probably the first of modern times. A macro-project is not a large one only by its unusual dimensions or by its technical perfection, but also because it is able to bring to all people more happiness, more friendship, more confidence in their capacity and future. To consider a macro-project as a global infrastructure solution is to be convinced that such a project is achieved by human beings and for them, and to act constantly under this conviction.

However, here I will not deal as much with the Suez Canal itself. Many people know that a plate of macaroni given daily to young Ismael, future Viceroy of Egypt, was the start of it all; that the Canal was the common undertaking of Ferdinand de Lesseps and the Egyptian viceroys, Ismael and Mohamed Said, against England; that it was solemnly inaugurated in 1869 and nationalized by Col. Nasser in 1956; that it links the Mediterranean Sea to the Red Sea; that it is 170 kilometers long and that ships pass through it in thirteen hours.

What seems to me much more worth telling is to emphasize that this macro-realization -- which the Suez Canal was -- created a waterway, but more than a waterway. It achieved the relief of men by machines. It gave a province to a desert. Such are the two items that I should like to consider here. Before that, however, I would like to briefly discuss de Lesseps' personality, because I think it concentrates some of the features of what can be imagined as the ideal macro-engineer.

## 1.1 THE CHIEF MACRO-ENGINEER: FERDINAND DE LESSEPS

Ferdinand de Lesseps was not an engineer; he had no technical knowledge. He was not an economist; he knew almost nothing about finance and banking. If he had been both or one of them, he would never have undertaken the enormous risk of such an enterprise: the technical difficulties undoubtedly exceeded the means available at that time; the conclusion of a cold analysis of its financial prospects would have been negative.

De Lesseps was an amateur; in fact, he was a diplomat, a retired diplomat. But he was an amateur with two invaluable gifts: will and imagination. "A bit of imagination," he said, "is not a bad leaven for the heavy dough of human affairs."

Imagination is probably the main quality required of a macro-engineer. Imagination made de Lesseps a macro-engineer. By it, he overcame the objections that reason and men oppose to any project of exceptional magnitude, and thanks to imagination, he was able to conceive the undertaking he planned and to be aware of all its implications. He met technical difficulties, financial difficulties, political obstacles. His answer was always the same: imagine what the Canal will be in ten years; imagine what it will bring to its shareholders, to Egypt, to the navies of the world; even to England, his main enemy.

And he was right. De Lesseps was not only an imaginative amateur, he was a generous amateur. He belonged to the generation of men and women who, in the middle of the 19th century, dreamed of a better society, who considered the workers' conditions of life, who brought to reality the first social welfare measures. The company established by de Lesseps was probably the first one to give, by a statutory provision, a part of its profits to all the members of its staff, including ordinary workers, and to build a system of family allowances, pensions, and social protection.

I wonder if generosity and disinterestedness are not, besides imagination, a defining feature of the ideal macro-engineer?

Ferdinand de Lesseps

Lastly, if de Lesseps was the leader, he never acted alone. He created around himself a team of men of ability and faith. The two Viceroys of Egypt (for Egypt was, at this time, under the theoretical authority of the Turkish government) supported the Canal project with intelligence and courage, for it was an important and dangerous undertaking for their country.

De Lesseps was helped by engineers of French birth but serving the Egyptian government: Linant Bey, Voisin Bey, and Mougel Bey. The contractors played an important part as well. And it should be mentioned that nearly all engineers and contractors had been trained in the *Ecole Polytechnique*, thus confirming again that a macro-project is necessarily linked to a team of people who share the same faith and the same dedicated enthusiasm.

## 1.2 THE BUILDING OF THE SUEZ CANAL

My first purpose is to show how the digging of the Suez Canal was a turning point in the management of important public works, and how, for men, it substituted machines.

The work site of the Suez Canal, in the open desert, was -- for the 19th century -- of an unprecedented magnitude, even if it was undertaken (by one of those historical coincidences) in a land whose inhabitants were renowned for gigantic temples and tombs. When the first studies took place, public works more or less ignored machines. Their basic element was manpower using shovels, picks, and wheelbarrows. In the second half of the 19th century, science began making an effort to theorize the efficiency of a man working in a building yard; however, the question of machines taking man's place had not yet been raised.

## 1.2.1   Manpower

Workers on important public works were, as they had always been, agricultural seasonal workers, ordinary prisoners, prisoners of war, or soldiers. For road and canal maintenance, some governments had a forced labor formula, days of work being a substitute for taxes of those unable to pay them in cash.

This type of work was a normal institution in Egypt. Irrigation is vital in this country, and the maintenance of a network of canals to distribute water from the Nile is a permanent necessity. Every village supplied a contingent of peasants who paid their tribute in kind, that is, by their labor.

Such was the formula adopted for the digging of this new, other canal, the Suez Canal. Egypt, Turkey, and England did not wish the *Compagnie* to use foreign workers, fearing they might stay afterwards in the country and constitute foreign enclaves. So, in the Act of Concession, it was agreed that the Egyptian government would provide all the workers the *Compagnie* might need.

Fellahs, called upon by the Egyptian authorities, came from their villages along the Nile and began to dig the Canal. In serving the *Compagnie*, they were well paid by Egyptian standards, and had decent food and lodgings. They worked on the site normally for one month, but many stayed on for a longer time. Twenty thousand were working on the Canal simultaneously. They were supplied with shovels and picks, as well as with ropes, driving wheels, ramps, and wheelbarrows. But they refused to use tools to which they were unaccustomed. The wheelbarrow, ascribed to a northern origin, had never been used in Egypt.

The first diggings were accomplished by fellahs wielding a shovel and a short pick; 35,000 such picks were imported in 1862. The sand was removed in baskets made of palm tree staple, and carried on mens' backs or by mules or camels.

Fig. 1.2 Local manpower working on the Suez Canal.

## 1.2.2    Manpower Replaced by Machines

Suddenly, in January 1864, everything changed: within a very few months, a revolution occurred. In 1863, the new Viceroy Ismael, pressed by England with the intent of ruining the *Compagnie*, told de Lesseps that he could no longer tolerate forced labor, which his government used. The Egyptian workers' contingents were progressively reduced, and in January 1864, following an arbitration by Napoleon III, the manpower hitherto supplied by the government of Egypt disappeared, after receiving compensation.

The engineers went to work immediately. They imagined and then designed machines specially planned for the Canal work, and then got in touch with manufacturers in England, France, and Belgium. Soon, in separate parts, the machines were shipped to Port-Saïd or to Alexandria, and by the freshwater canals or on camel-back, and subsequently delivered to the appropriate Canal work sites.

One year after the arbitration award by Emperor Napoleon III, the canal works had jumped over centuries: where men whose methods closely resembled those of their forbears who built the pyramids, where their tools, their movements, their attitudes had not changed through 7,000 years, now steampower was operative.

A hundred different steam machines came into action along the Canal route and fewer than 2,000 workers were necessary. What were those machines?

The seven volumes of the Voisin Bey report on the Canal's construction, which recount its progress month by month and set forth the original texts of the agreements with the various contractors, affords us a precise idea of the dynamics of a mechanized public works project in the middle of the 19th century. Our archives retain contemporary plans and photographs. The pictures are touching, and remind one of the drawings in 19th century novels: big chimneys, huge metallic structures. But the machines worked and, with good mechanics, worked very well.

The machines were mainly dredges. The largest part of the Canal construction was done under water: the opening of two ports, Port-Saïd and Suez, and the digging of the Canal through Lake Menzaleh and through Bitter Lakes. A special technique, when and where applicable, was used to extend underwater operations. A first trench was dug by hand or by an excavator; the freshwater canal was connected with the trench, and the water coming from the Nile filled it; the dredges could then dig to the required depth.

Dredges were of two types: first came the classic small dredges 20 meters long and 7 meters wide, 18 horsepower machines. Soon one of the contractors designed a new, more powerful type of dredge. It was 30 meters long, with a 40 horsepower machine capable of dredging at a depth of 8 meters. The primary novel feature of the new machines was a corridor 100 meters in length along which sand was pumped and ejected onto the desert. Twenty of those dredges were manufactured in 1864, and twenty more soon followed.

For dry works, the same contractor built an excavator -- a sort of earth dredge moving on rails, with two steam engines and a chain of buckets. Sixteen were used on the Canal.

**Figure 1.3    Steam-operated dipper dredge with long boom to deposit excavated sand on embankment.**

Another new machine, specially designed for the Canal, was an elevator which was used when the banks were too high, a sort of moving bridge which elevated tanks full of sand and carried them over the banks to the desert where the sand was discharged.

Locomotives began pulling trucks, which were originally pulled by camels or mules. Thirty-seven "hopper barges" were built in Glasgow, Scotland. Filled with sand or silt, they could move their cargo to the open sea where it was discharged.

Cranes, pumps, tugboats -- all steam-powered -- were busy on the sites.

The hardest rocks were handled by explosive powder, most noticeably on the day before the Canal was solemnly inaugurated, because several blocks still remained which made impossible the transit of some of the 70 ships waiting at Port-Saïd.

The ultimate resource was powder, and it worked. Concrete was used to build Port-Saïd harbor; in the ancient times, blocks of stone had been carried, with great difficulty, from Alexandria and the Greek Islands; afterwards, the piers were constructed with concrete made at the site. The first concrete lighthouse in the world was built at Port-Saïd. Thus, the future met the past in the land of the Pharaohs.

Steampower relieved manpower. Steam was the motive force that rescued the project and paced the construction of the Canal. As with its building, the opening of the Canal took place at the hinge of two epochs: in 1858, when the Concession was given to de Lesseps, most ships crossing the seas were wind-powered sailing craft; twenty years later, they had practically disappeared. Without steampower and steamships, the Canal would not have been used. Building the Canal was a gamble on the timely arrival of modernity -- and steam won the bet. Today, steam is no longer present on the canal banks.


## 1.3 THE SUEZ CANAL TODAY

Today, the dredges, more powerful, more modern in many respects, remain the beating heart of the Suez Canal. Night and day, without stopping, they are at work along its 170 kilometers. Ships in transit make destructive waves which cause the banks to erode and crumble. Sand must constantly be dredged and discharged.

Despite such problems, the Canal is not only maintained, it is improved and made deeper: from 8 meters in 1869; to 13 meters in 1939; 15 meters in 1955; and 20 meters today. In some sections, depths of 60 to 120 meters have been achieved.

The dredges have been there from the beginning, and the Canal requires continuous restoration.

## 1.4 THE IMPACT OF THE CANAL ON MAN AND HIS SURROUNDINGS

Now I will proceed to the second part of this Suez story. In the first half, I tried to show how a macro-project can renew the conditions of man's work. At this point, I will tell how it can modify his surroundings.

The Suez Canal is not only a waterway, it is a province conquered from the desert. Certainly, it can be said that American towns, all American states, were born from "desert", and that this is a normal circumstance for the inhabitants of this part of the planet. Yet there is a difference, a fundamental difference. Towns usually derive from geography; they are built on the banks of a river or lake, set on natural bays or passages.

The Isthmus province is born against geography, in one of the most hostile natural settings; it developed from the Canal, that is, from an individual's macro-project. In 1858, the Isthmus was a dead desert. Men passed through; no one ever decided to stay. The Hebrews fled the Army of Pharaoh. Egyptian, Hittite, Greek, Roman, Arabic armies crossed the sands, but did not linger to establish settlements. Caravans joining the countries around the Mediterranean Sea to the lands of the East and the Far East followed the desert tracks to or from Sinai, Palestine, and Syria. But between the last village of the Nile Delta and the Gulf of Aquaba, no oasis, no village, not even a nomad camp was to be met, except a small fishing port on the Red Sea: Suez. The North Coast was low and inhospitable; Lake Menzaleh, to the south was more a marsh than a lake. Afterward, the desert rose to some 120 meters, to slope down again to Timsah and Bitter Lakes depressions, then desert again and the Red Sea shore.

Today, we are familiar with bases of operations where technicians and workers engaged in a big project, or soldiers in a military campaign, live comfortably in a desert or on an ice pack. But in 1859 it had never been done. Two years later, de Lesseps wrote:

We had to build or buy enormous quantities of equipment; we had to create, on the Mediterranean Sea, a town of four thousand inhabitants, with workshops and machines of all sorts; we had to bring the Nile to the desert; we had to carry all our equipment and all our supplies; our organization is now completed and can sustain, feed and give working tools to a pacific army of forty thousand workers.

### 1.4.1   Building Cities

The digging began on the north part of the Isthmus because sources of supplies of all nature were nearer. But before the first shovelful of sand was removed, two major problems had to be solved. The *Compagnie* encountered exactly the same difficulty General Eisenhower had to overcome on D-Day: no harbor was available for the ships coming with men and supplies. The only solution was to create one. Port-Saïd was created *ex nihilo*.

### Port-Saïd

Piers had to be built. First, stones from quarries near Alexandria or even in the Greek Islands were used, but the cost of moving them became prohibitive. The engineers decided, probably for the first time, to make artificial blocks of lime mortar. Each block weighed 22 tons and was 12 cubic yards in size. A total of 30,000 blocks were used.

Quays, wharfs, progressively appeared. A channel was dredged, used by ships of increasing tonnage as the channel became deeper and wider, bringing in more supplies, more ponderous machines. Dredges and barges came, under their own power, from Europe.

In 1859, the population of Port-Saïd was 150 inhabitants; two years later, there were 4,000.

The second problem, perhaps even more vital than the creation of a harbor, was a supply of drinking water. Egypt is a gift of the Nile River. The cities and fields of the Isthmus which exist today are, also, a gift of the Nile. But the Nile had to be brought to the Isthmus. Before building a maritime Canal, the *Compagnie* had to dig a freshwater canal.

A reliable supply of fresh water was assured and achieved, at least during the first year, by transporting water from the Nile in goatskin containers on camel-back or on small boats through Lake Menzaleh. Subsequently, three steam distilling machines were imported so that water from Lake Menzaleh and from two half-salt wells discovered in the south, could be used. The digging of a freshwater canal was pushed.

From the Nile, near Cairo, an irrigation canal brought water to a big agricultural property. This canal was enlarged and deepened, and was extended to a point situated roughly in the middle of the future maritime canal, where Lake Timsah and the town of Ismailia were to appear. From this point, two branches were dug, one south to Suez, the other north to Port-Saïd. By the beginning of 1861, fresh water from the Nile watered the sands of the desert.

The first building of the future town of Ismailia was a water station for pumping, distilling, and distributing water from the freshwater canal along the north part of the canal and Port-Saïd. Four steam engines were at work. Tanks were disposed along the work sites. In 1864, street fountains appeared along what were later to become Port-Saïd streets.

Three main towns developed along the Canal. Between them, some technical stations, and along the sweetwater canal and irrigation ditches, cultures. In Egypt, where the water comes, immediately palm trees, vegetables, sugar cane and cotton began to appear.

Port-Saïd was, at the beginning, a harbor, and remains a harbor today. But an important city developed around the maritime setting. Sand and stone dredged to create the harbor and its channel were discharged onto the shore to make a large platform where the town was built. Ships coming from the north to enter the Canal, or leaving the Canal from the south, stopped in Port-Saïd to coal, to refill their water tanks, to buy fresh food, or seek repairs. Passengers disembarked, and some went to Cairo and the Pyramids for a day. A night crossing was not immediately possible, so ships stopped overnight at Port-Saïd, and soon the city became a specialized and prosperous site for sailors.

Today refueling is not always necessary. Many ships wait in the harbor just long enough to take their places in the south-bound convoy, or leaving the Canal to the north, they don't even stop.

The activity in Port-Saïd gradually became independent of the Canal and today the city is one of the most important ports in the Mediterranean Sea, and the principal port of Egypt. Industry and commerce are active, and a free zone has been created.

Port-Saïd is not an Arabic town; it has no *vieux quartier* like Cairo's or Alexandria. It is a Mediterranean city which keeps the imprint of the *Compagnie*: well-built, with wide avenues, modern hotels, mosques, churches, and schools. Shipping companies throughout the world maintain offices in Port-Saïd and leading commercial, industrial and financial companies have representatives. The Suez Canal Authority has its local office, as the *Compagnie* had, in a large building along the Suez Canal entrance.

Earlier, a statue of Ferdinand de Lesseps, nine meters high, looked down on the harbor. It was blown up in 1956, but repaired in the early 1990s and there is a good possibility that it will soon be re-erected on the bank of the Canal, to salute transiting ships.

## Port Fouad

By 1920, Port-Saïd was already too small. A new town was built by the *Compagnie* on the other bank of the Canal, as a sort of Port-Saïd suburb. Its name is Port Fouad, in honor of King Fouad who solemnly inaugurated it. Port Fouad is mainly a dockyard where every tool used on the Canal could be built or repaired. Around the dockyards, houses were built, schools opened, and large parks and playgrounds were created. Ferryboat service links Port Fouad to Port-Saïd.

South-bound travellers can use a large new motorway stretching from Port-Saïd to Ismailia. The water route goes along the sweetwater canal, the railway line, and the old strategic road where British troops were on guard until 1952, and which is used today by the Egyptian Army and the Canal agents. Behind trees, the sand dunes forming the banks of the Canal appear.

Halfway from Port-Saïd to Ismailia, a branch of the road goes east, by a tunnel under the Canal, to Sinai, Israel, and Jordan. The tunnel just recently opened; earlier, the Canal was crossed by ferries and by one turning-bridge, which was especially dangerous for transiting ships.

## Ismailia

Ismailia, like Port-Saïd, is entirely a creation of the *Compagnie*, although quite the opposite character of Port-Saïd. It is a residential resort along Timsah Lake. The Canal waters transformed a desert depression into a large lake where transiting convoys from north and south can cross and where Ismailia inhabitants can swim and sail. The *Compagnie* opened the streets, set up gardens and parks, built schools, churches, and mosques, and organized all the town services. Later, when the Egyptian government established a local administration, all these services were transferred to the local government.

Ismailia is a large garden with trees, lawns, and flowers. The streets are large, the houses built for the Canal agents comfortable. The office buildings of the *Compagnie* are unchanged from the beginning of this century: low, made of brick, and rather old-fashioned.

Today a high building, white and facing the lake, is the heart of the Suez Canal Authority. It is set in a park, against the water and sky, and from the offices one can imagine oneself in a luxury hotel more easily than in the operational center of the Canal. In fact, in Ismailia, one plays tennis and golf; clubs, inspired by a certain colonial way of life, are still active along the lake, as tankers and container-ships pass by far away.

The hospital built by the *Compagnie* and managed under its authority until 1956 remains one of the most modern in the Near East. The water station has been continuously modernized and distributes fresh water to Ismailia and its surroundings. Many churches remain, as well as schools of all origins. On Chevalier Island (today Peace Island), Anwar el Sadat met Menachim Begin. President Mubarak has a residential house there, overlooking the Canal. The residence where de Lesseps lived and where so many "personalities" spent their stay in Ismailia, is still in use, perfectly well-maintained. Ismailia is and will remain the "brain" of the Canal.

But it is creating its own independent life as well. On the other side of Lake Timsah, residential flats are built for civil servants coming on weekend or holidays from Cairo. Ismailia can become, too, a tourist turntable to Cairo (west), Sinai (east), and the Red Sea (south).


### Suez

At the end, on the Red Sea, is Suez. It is probably the town which the Canal has transformed the least, but it retains a special symbolic value. Suez has always been there, a small fishing village, from the dawn of history. It gave its name to the Isthmus and to the Canal. It gave its name, too, to the company that created the Canal and managed it for nearly a century.

When one spoke, and speaks today, of that company, of its shares, of its dividends, one uses the phrase, "the Suez," not knowing exactly where Suez is situated on the map. After the nationalization of the Canal in 1956, negotiations took place between the Egyptian government and the *Compagnie* to discuss compensation and to settle the questions of the belongings of the company in and outside Egypt. One of those belongings was certainly the name, Suez. To decide whether the *Compagnie* could keep and use it was discussed for several days.

I remember one evening, in Cairo, we had been arguing with the Egyptian delegation, in a rather confusing oriental method, and without any result. As we were about to adjourn the meeting and go back to our hotel, one of Colonel Nasser's assistants came in and told us that the colonel would be there in a minute. He came in, and we all were silent. He just told us, "I keep the Canal; Suez, I give it to you."

That's how, today, the second de Lesseps child after the Canal, that is, the company which became after 1956 an investment company, is named *Compagnie de Suez*.

Suez is a port where the Canal begins, or ends, for transiting ships. But the town is six kilometers from the port. Port equipment itself is situated at the end of an artificial peninsula in a new town called Port Tewfik.  Transiting ships generally wait in the bay, which is spacious and calm, before taking their place in the convoy north-bound, or leaving the Canal for the Red Sea.

Suez is, in fact, less dependent than the other towns on Canal activity.  It has its own industries and today is one of the centers of the Egyptian petroleum industry.  Although heavily bombed, it has now been fully reconstructed with Saudi Arabian help.


## 1.5 CONCLUSION

The vision I have attempted to sketch, of what was in so many respects, a macro-adventure, ends rather like the last chapter of a guidebook, describing a commonplace town anywhere in the world.

Indeed, revolutionary projects, when achieved, are gradually integrated into everyday life and appear as something "normal".

Thus, man's imagination can move forward, giving birth to new macro-realizations which, as time goes on, will appear as part of our everyday routine.  If, one day, you should travel in Egypt, not far from Ismailia, stop on the road bank.  You'll see, first, fields cultivated by peasants, donkeys passing by, cows drinking.  Then, just a bit further, suddenly as if by magic, huge 200,000 ton tankers, container ships as large as towns, will appear as if they were travelling through the fields.  You'll remember the Canal and all that humanity invested to open the earth to those vessels, coming from all over the world.

Then, at last, turn around.  The desert is commencing just on the other side of the road.  Then you'll think that, on our planet, many deserts are still to be opened and fertilized by macro-engineers with great imagination.

# 2

# Operation Mulberry: A Floating Transportable Harbor for World War II Normandy Invasion

**John W. Landis**
*Stone & Webster Engineering Corporation, Boston, Massachusetts*

The winter and spring of 1944 witnessed the most remarkable and intensive build-up of military might in history. The locale was southern England. The reason for the build-up, of course, was the impending invasion of France.

Those of us who were privileged (and I say "privileged" only because I view the situation now from the vantage point of almost 50 years) to participate in the build-up and subsequent invasion can never forget certain aspects of that particular episode in modern history.

One thing that undoubtedly sticks in the memory of every American who was there was the dedication of the British citizenry to the survival of their country. I have the uneasy feeling today that no large group of people will ever again demonstrate such devotion, loyalty, and sacrifice. From wretchedly damp beds in the subways of London, from squalid pads in makeshift shelters put together from the rubble of cities like Portsmouth, from board cots in sentry huts along the coast, from cold couches in the rural villages, they rose each day with an implacable determination to do what had to be done to get ready for D-Day. Despite the awful physical and mental damage wreaked by the Nazi air raids and propaganda barrages, they showed little emotion and absolutely no panic. They made wry jokes about their plight but they indulged in no self-pity. They were steadfast, cheerful, and strong, as I hope the people of this country will be if missiles from some distant foe ever rain down on our territory.

The results of their labors were as impressive as <u>they</u> were. Every highway, road and country lane feeding into the many port cities of southern England was neatly lined with the mobile weapons and military equipment of that era. The bombed-out areas around the docks were cleared of debris. Fenced-in warehouses and acres of quonset huts were stuffed with supplies. Trim ships of all sizes nestled alongside the wharves. Miraculously, the railroads were relatively undamaged and the trains were running on schedule. Electric power was available almost everywhere. Telephone and telegraph service was reasonably good. Other essential services were being provided.

Meanwhile, in the various military headquarters and operational facilities scattered throughout Great Britain, plans were being developed and preparations were being made for the most daring and massive attack ever mounted by one power against another. The code name for the invasion was Operation Overlord. Those who knew about it were sworn to absolute secrecy -- "bigoted" was the British term. Some of the pledges and penalties involved in this procedure would make interesting reading but I shall bypass that subject because I am not certain that it has been officially declassified.

I was "bigoted" in April 1944. Perhaps I should back up a bit and tell you how I arrived at this point in my short naval career.

When my commission came through in January 1943 the Navy apparently was looking for physicists and engineers skilled in influence mines-countermeasures. Influence mines were those detonated by magnetic or acoustic impulses. The counter-measures used at that time included "sweeping" with artificially created magnetic or acoustic fields, eliminating certain noise frequencies emitted by ships, removing the permanent magnetism pounded into ships during construction, and counteracting the magnetism induced in ships by the earth's magnetic field. My first orders sent me directly to the Naval Mine Warfare School in Yorktown, Virginia. After ten weeks of study there, I was assigned to the Boston Navy Yard for two months of practical training, and then transferred to the Bureau of Ordnance in Washington, D.C. for additional training. In July I reported to an advance base unit at the Norfolk Naval Operating Base preparatory to going overseas as the commanding officer of a small-mine warfare unit.

My unit was shipped to Great Britain in January, 1944.  We crossed the Atlantic on the Queen Mary, unescorted, zig-zagging at top speed all the way to elude German U-boats.  There were 17,000 men and women on board, and it took two full days to unload in the Firth of Clyde.

As soon as we landed at the naval base (Base Two) in Rosneath, Scotland, it became clear that the need for American influence mines-countermeasures officers was not nearly as great as the strategists in Washington had thought.  The British Navy had trained more than enough such officers to satisfy the requirements of Supreme Headquarters.  So, although I still had to test my equipment and put my crew through degaussing, deperming, and mine-disposal maneuvers, I was gradually given other assignments, including inspecting British port operations.

This period of "diversified education" ended when I was ordered to Portland and London in mid-April for temporary duty of an undisclosed nature.  I had no idea what I was getting into -- but I recall a strong gut feeling that a new phase of my naval career was opening up.   The formal report I made to my commanding officer upon my return to Base Two tells it all -- although not very elegantly:

1.  From 15 April to 17 April: attached to U.S. Navy Tug Depot, Portland, England.  Gleaned a smattering of towing techniques to be utilized during invasion.

2.  From 17 April to 22 April: attached to C.T.G. 127.1 in London.  Had access to complete plans of invasion.  Studied these with end in view of becoming general dispatching officer for U.S. port material.  Will be aide to Commander Kitcat, Royal Navy, aboard the Queen of Thanet.

"Had access to complete plans of invasion" was my juvenile and rather cryptic way of describing the "bigoting" process.  Looking back, I realize that what I went through in that short week was a turning point in my life.

After being sworn to an ultimate degree of secrecy (and I must confess that the burden of what I was told weighed more heavily on me during the next two months than anything I have ever experienced, before or since), I was taken to a heavily guarded underground complex in London where a complete model of the beaches to be invaded at the outset of Operation Overload was explained to me.  I was given dates, names, and locations of key targets; outlines of the planned operations; numbers of ships, personnel and weapons involved; and even a short discourse on strategy.

During my stay in London I was completely miserable living with this knowledge.  Nothing else seemed of much consequence -- that is, until I discovered that the seemingly solid ceiling of my top-floor room in an old apartment house was really a glass skylight painted black.  Then I found a queer kind of relief in switching my worrying back and forth between fear of revealing vital secrets and fear of being lacerated by flying glass from the bombings that occurred every night.

One of the postulates of the Allied High Command planning the Normandy invasion was that the Nazis would recognize the critical importance of the harbors in northern France and fortify them completely. The Nazis, it was believed, would reason that if the Allies could be denied the use of these ports for a protracted period after the initial landings, the invasion would falter and be crushed for lack of equipment and supplies. This indeed proved to be their thinking and it was sound except for one thing -- it did not take British and American ingenuity into account.

Early in 1943, British Army personnel assigned to the task of developing alternate methods of bringing ashore the vast amounts of materiel required to support a full-scale military operation on the European mainland decided that the best method would be by means of conventional shipping protected by artificial harbors erected at the main beachheads. The big question was: Would such a huge and complex project be feasible under battle conditions?

Spurred on by Prime Minister Churchill and President Roosevelt, to whom the concept had been revealed, a team of British and American engineers designed a harbor that could be built in pieces in Great Britain, towed across the English Channel during the first days of the invasion, and erected at any one of several possible beachheads. Supreme Headquarters agonized over the plan, but eventually decided to adopt it. A definitive scheme for the entire operation was developed and submitted to Churchill and Roosevelt at the Quebec Conference in August 1943. They approved it, and the preliminary work on fabrication of the basic components of the harbors that some brave but unknown military personnel had authorized immediately burst into a full-fledged, all-out effort.

The plan, which was given the code name Operation Mulberry, can be summarized as follows:

1. Build two artificial harbors -- one for the proposed American beachheads near St. Laurent sur Mer, the other for the proposed British beachheads near Arromanches -- each as large as the man-made port of Dover.

2. Each harbor would consist of an outer roadstead sheltered by floating steel breakwater units called "bombardons", an inner roadstead sheltered by a line of blockships sunk end to end and a line of huge concrete caissons called "phoenixes", and three 3000-foot floating steel-and-concrete causeways called "whales" connected at their outer ends to "spud" pierheads. ("Spud" pierheads are steel docks that ride up and down with the tide on adjustable vertical legs driven deep into the sea floor.)

3. Freighters, tankers and other large ships were to be anchored in the outer roadstead; LSTs, smaller types of landing craft, coasters, barges, tugs and other service vessels to be anchored in the inner roadstead.

4. The blockships would be steamed to position and sunk first; the phoenixes then towed to position and sunk next; the bombardons to be anchored as soon as their correct positions outside the line of phoenixes could be determined. Concurrently with the latter two operations, the causeway sections would be hauled in and moored from shore out, with the pierheads arriving in time to be hooked on their outer ends.

5. The pierheads would be capable of taking tanks and heavy trucks from both the bows and the decks of the LSTs.

6. Traffic on the heavy causeway (40-ton capacity) was to be restricted to Sherman tanks whenever necessary.

7. Traffic on the other two causeways (25-ton capacity) would be restricted to trucks and lighter vehicles at all times.

Rather than give detailed descriptions of the various components and pieces of equipment involved in this project, the illustrations that follow convey far more than long strings of words.

**Figure 2.1 Floating breakwater unit, called "bombardon".**
*(Its cross-section, mostly below water, is cruciform in shape. The counter-acting subsurface forces created when a wave hits the top segment attenuate the wave's energy.)*

**Figure 2.2  Line of blockships sunk to create an initial breakwater.**
*(Sunk about 5/8ths of a mile out from Omaha Beach, these blockships were placed neatly end to end under very difficult weather conditions and heavy Nazi artillery fire.)*

**Figure 2.3     6000 ton "phoenix" being maneuvered into position by harbor tugs.**

**Figure 2.4  Cluster of "spud" pierheads being assembled into a facility suitable for unloading tanks, trucks, bulldozers and other heavy vehicles.**

**Figure 2.5  Causeway, about 3,000 feet long, consisting of several pontoon bridging units, or "whales", terminating in a shore ramp.**

When I returned to Base Two from London on April 24, 1944, I learned that within a day or two I would be receiving official orders to proceed to duty on the Queen of Thanet at Sheerness at the mouth of the Thames. I only had time to collect a few belongings, inventory my countermeasures equipment, turn over the command of my crew, seal my sea chest and write to my wife before I took a night train south again. I remember well my tingling anticipation of this new assignment. I also remember well the great letdown I suffered when I saw the Queen of Thanet. It was an ancient paddle-wheel steamer, built in 1890!

The British crew and staff aboard the Queen, however, made up for her physical deficiencies. They were an exceptionally alert and capable group -- one which made me run, both literally and figuratively, to hold my own. I never did fathom their fluid rules for card games, however.

My official position was "American Liaison Officer", but in reality I did whatever the commanding officer, Commander Kitcat, needed done -- as did everyone else. It was the most flexible working arrangement of my naval career -- and the most efficient. The job of getting the vast array of equipment needed for the two artificial harbors ready for towing across the channel and then seeing that it actually got to the far shore during the invasion was one which could never have been done by so small a group under standard operating procedures.

**Figure 2.6    Mulberry dispatching officers on the bridge of the Queen of Thanet**
*(left to right: Commander J.P. Kitcat, RN; Lt. R.E.J. Browne, RNVR; the author; Lt. Commander C.W. Hancock, RNR).*

We sailed out of Sheerness in a small convoy shortly after I boarded ship. The other vessels in the convoy were a tug and a coastal tanker. Making all of four knots, we headed through the Strait of Dover, hugging the shoreline as closely as we could. Sailing along a land mass that you know should be dotted with lights, and finding it completely blacked out -- you may understand when I describe my reaction as an eerie feeling.

I recall being reasonably relaxed during the first stages of the voyage, however. The main reason for this was simply that my British colleagues had withheld from me some recently acquired information about intense E-Boat activity out of Calais. I was the perfect example of the old adage, "Ignorance is bliss".

Their well-intentioned deception came to an end, however, when we were passing South Foreland. Suddenly several shore batteries opened fire on a fast-moving target (pinpointed by the brilliant red tracers) between us and the shore. Then I was told about the E-boat activity -- and my comfortable drowsiness gave way to skin-crawling wakefulness. The Chief Engineer immediately shut down our engines; we were careful to move silently and to speak in whispers. I put on my lifejacket, fully expecting to have to make good use of it. Commander Kitcat insisted that we wear our helmets and cautioned us to keep them on in the water.

The firing continued for about 15 minutes; then it stopped abruptly. We lay still in the water for another hour or so, seeing nothing, hearing nothing. The longer we waited the worse the tension became, because all of us knew that the crucial moment would be when we started our engines again.

Finally Commander Kitcat gave the signal to move. The engines began their throbbing beat -- horribly loud to our straining ears; the paddle wheels began their churning and slapping. No torpedoes. We went to bed and awoke to a bright sunny day off Dungeness.

How long it took us to get to our destination -- which was Selsey Bill, the point of the peninsula just east of the Isle of Wight -- I do not remember. But I do remember my surprise at what we saw when we arrived. Parked in the gravel-bottomed shallow water of that area was a large portion of the massive array of components needed for the Mulberry harbors. There were so many of them and they were scattered so haphazardly over such a vast area that I wondered how we would ever organize the dispatching operation.

Another of my first reactions was that the Nazis could not help but know what we were up to if they had any aerial pictures of that scene, and I was sure that they had. I learned later that they thought the phoenixes were mobile anti-aircraft stations to be parked at strategic locations along our beachheads.

What they thought the pierheads and causeway units were I never found out. No bombardons were moored in the Selsey Bill area.

**Figure 2.7     Mulberry storage, testing and dispatching depot off Selsey Bill.**
*(Location of the various components, primarily "phoenixes" and "spud" pierheads which needed the most maintenance and rehearsal, was dictated by sequence of receipt, ocean-floor characteristics, weather, availability of harbor tugs, and many other factors.)*

Around-the-clock work on the major components for the Mulberry harbors was started immediately after the Quebec Conference. Shipyards and steel-fabricating facilities throughout Great Britain were pressed into service. Activity quickly mounted to a fever pitch and continued at that level throughout the early months of 1944, so that by the time we arrived at Selsey Bill, about 90% of the components had been completed and stored there and in the other areas selected on the basis of good seabed characteristics and good correlation with the logistics of the Overlord plan. Selsey Bill was the largest of these parking areas.

Our first job at Selsey Bill was to draw up a detailed master plan for checking and maintaining the various components prior to the invasion, which was then set for June 5th, and for getting them across the Channel and in place at the beachheads. In carrying out this assignment we worked closely with the British Royal Engineers, who had been selected to erect the British harbor, with the U.S. Seabees, who had been selected to build the American harbor, and of course with Supreme Headquarters.

Next, we inspected, repaired, and tested the components. Then we practiced with them incessantly, until we could literally handle them in the dark. We had to be prepared for any eventuality because we were told that the success of the invasion depended on our getting the harbors into operation on schedule -- that is, by D-Day plus 10.

The harbor components contained some rather complex mechanisms -- valves at the bottoms of the phoenixes, engines and gears to move the long legs of the pierheads up and down, and anti-aircraft guns, to mention just a few. We had more than the usual trouble with these mechanisms because a substantial number of them had been manufactured and installed in great haste by contractors poorly equipped to do such work. The Royal Engineers and Seabees had to finish or repair many components after they were received in the parking areas. In addition, they had to train dispatching, riding, sinking, erecting, operating, and maintenance teams. They did a tremendous job. The Seabee unit assigned to this project, incidentally, was the 108th, commanded by Commander E.T. Collier.

One challenge deserving special mention is that of sinking the phoenixes without capsizing them. These units, which were about 200 feet long, 50 feet wide, and 60 feet deep, weighed 6000 tons, and were compartmented for strength. Thus, if the valves at the bottom were not opened in such a way that water would flow into all compartments evenly the unit would list sharply and, being flat-bottomed and hydrodynamically unstable, possibly turn over if the tides and wave action were wrong.

Handling these massive pieces of the Mulberry harbors required a large fleet of harbor tugs, sea-going tugs, ducks, and service boats. Also, salvage vessels were needed to pump the water out of the parked phoenixes after their valves had been closed. The procedure was to close the valves, pump out all compartments evenly, gradually raise the unit to the desired draft, and tie it to a couple of harbor tugs. The harbor tugs would then move it gingerly out of the parking area and hook it to the end of a 1500-foot cable from a sea-going tug. The seagoing tug would tow it across the channel to France.

After we had progressed far enough with our checking and testing at Selsey Bill, we made several trips up The Solent and to the other nearby Mulberry areas to coordinate the entire operation. Whenever we docked near a U.S. base on these voyages it was my custom to contact the mess officer at the base and scrounge a few cans of fruit and vegetable juices to augment the rather austere menu aboard the Queen. My British colleagues appreciated this very much and, I am afraid, treasured the juices to the point of absurdity. On one occasion they insisted on saving some tomato juice for a week to have it with a birthday dinner for the Chief Engineer. When we finally drank the juice, conditions in the galley being what they were, it had black lumps in it. Though I drank my portion down

**Figure 2.8**     **"Phoenix" being towed across the English Channel by a seagoing tug.**

without changing expression, as did everyone else, I could not resist investigating the situation. That evening I sneaked into the galley and discovered that the tomato juice tin was full of cockroaches.

About the middle of May, after we had crystallized our needs for harbor tugs at Selsey Bill, the British officer who was second in command on the Queen of Thanet, Lieutenant Commander Hancock, and I journeyed to Southampton to take over some U.S. tugs that Commander Kitcat had requisitioned -- a fleet of eleven STs. Since I was the only U.S. officer on the headquarters staff, I was to command the fleet. No one knew -- and for some reason I could not bring myself to tell anyone -- that I had never had a minute's training in the operation of any kind of vessel, much less a tug. I reasoned, I suppose, that the individual tug captains would know all there was to know about their vessels.

We arrived in Southampton late in the afternoon and had to argue with the U.S. authorities there all night to get the tugs that we wanted -- and desperately needed. I do not remember what the bone of contention was; I recall only that it was a trivial matter which roused my ire to the vaporizing point.

My mood was therefore far from pleasant on the next morning when I ordered the merchant marine crews to get ready to sail. You can imagine my reaction when one of the captains refused to go unless he received triple pay for working in a battle area. We had a few choice words and finally I told him to pack his belongings and get off the ship -- that I would take over.  One of the biggest shocks I ever received in my life was his quiet acceptance of my ultimatum.  To this day I do not know whether he was deliberately calling my bluff or whether he really was afraid to go to the Channel area.

In any event, I suddenly found myself in command of a tug and a suspicious crew of six experienced seamen.  The other tugs were already shoving off so I knew I had to quickly show some authority.  I took the bridge with all the confidence I could muster and as soon as I spotted a member of the crew starting to do what appeared to be an essential task I barked out an order to him, instructing him to do what he had already begun.  Fortunately, this maneuver gave the crew the impression that I knew at least the rudiments of seamanship, and they went about their duties with at least a semblance of discipline.

I was not out of the woods entirely, however.  By the time we reached the ship channel in The Solent we were well behind the other tugs and I had trouble with the channel markings.  Somehow I strayed inshore of one of the buoys and before I realized what had occurred we started scraping the bottom.  My heart sank.  I could visualize the court martial coming up in all its gory detail.  Then I remembered a piece of advice the famed salvage expert, Captain Edward Ellsberg, who was assigned to Mulberry at Selsey Bill, had given me:  a tug has enough power to plow through shallow sandbars.  So I signalled full-speed ahead and we shuddered on through.  The old saying, "It's frightening to realize how much one's future career hinges on little things," was no longer trite to me.

During the few weeks I had that little tug in my command I became quite attached to it -- and, with the help of an understanding crew, learned a great deal about operating it.  As a matter of fact, it won me one of the most cherished accolades of my naval service.  During the last practice maneuver prior to the invasion, the companion tug working with me ln repositioning one of the 6000-ton phoenixes lost its power and I had to handle the massive unit alone.  Tide and wind conditions were just right and I was able to bring the unit alongside the waiting salvage ship very gently.  As we closed the last few inches, a member of the ship's crew held an egg in the air and then with a flourish placed it between the gigantic masses coming together.  Of course it was crushed, but I learned later that this is the British seaman's way of saying "well done" to a tug captain.

On June 3, we started preparing our first tows for the invasion.  The seagoing tugs were on hand and all components were in top shape.  On June 4, we received word that the weather forecast for the 5th, General Eisenhower's first choice for D-Day, was bad -- heavy seas and high winds.  Orders came through to postpone all operations 24 hours.  Late on the 5th, after the meteorologists had

Figure 2.9  Harbor tug temporarily commanded by the author.

Figure 2.10  Crew of a harbor tug during a rare moment of relaxation.

assured Supreme Headquarters that winds and waves would abate a bit on the 6th, confirmation of the big decision was received: Go ahead!

We worked all that night, all the next day (D-Day) and through most of the next night.  Again I marveled at the fortitude and stamina of my British companions.  They kept going until they were hollow-eyed, bristling with beards, hoarse and unable to stand erectly.  But they never lost their concentration on the mission to be accomplished or their ability to make sound decisions.  From what I learned later from observers like Captain Ellsberg, the Americans involved in that frenetic activity also performed admirably.

Standing on the deck of the ancient Queen of Thanet, I felt like a carryover from a bygone age watching new and wondrous technologies being implemented to accomplish miracles that were beyond my ken.  The orderly mammoth stream of bombers overhead was in stark contrast to the frenzied and complicated activity on the shore and in the waters of the Mulberry park.  Despite the many weeks of careful planning and practice, nothing seemed to be organized -- but somehow the job was being done.  Every single Mulberry component went out of Selsey Bill on schedule.

On D-Day plus 2, I boarded one of the second wave of seagoing tugs taking the phoenix units to Normandy and made the 30-hour journey to Omaha Beach to see what was going on.

What I saw disheartened me.  Despite the heavy aerial attack started by the endless waves of bombers the night of the 5th and continued without cessation through the following three days, many of our forces were still pinned down in the beach area, immobilized by a combination of land mines and shell fire. Although naval guns had finally cleaned the Nazi artillery out of its emplacements along the top of the formidable bluff overlooking the beach, the toll the enemy guns had taken in men, materiel and ships was evident all along the water's edge. There were dozens of blackened tanks, damaged landing craft, wrecked trucks, demolished jeeps and punctured ducks scattered throughout a maze of barbed wire, displaced beach obstacles, discarded weapons and seaweed as far as I could see.  Combat engineers and Seabees were valiantly trying to clean portions of the beach with trucks, tractors and bulldozers, but I could not detect much progress. The most distressing sight, however, was that of the burial parties collecting bodies in big tarps and rolling them into makeshift graves.

Progress on the harbor was good, however.  The twenty or so blockships, headed by the old British battleship CENTURION, had been sunk in perfect position without premature loss of a single one (although one did hit a mine just as it was being maneuvered into position).  The huge armada of components sent over from Selsey Bill and other parking areas had arrived virtually intact as far as I could tell, and the Seabees were carrying out the Herculean task of anchoring the bombardons and sinking the phoenixes with excellent results despite the heavy swells that continued to plague them.  Already the first causeway unit was being tied to the shore and the first pierhead was being towed into the area.

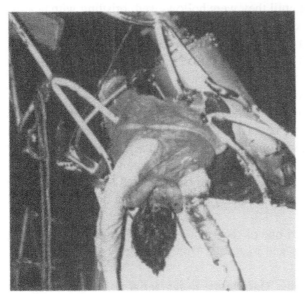

**Fig. 2.11**      **One of many U.S. ships and small craft that suffered direct hits by Nazi artillery.**

**Fig. 2.12**      **Some of the wreckage resulting from the first attacks on Omaha Beach.**

**Figure 2.13  Early casualty in Normandy.**

**Figure 2.14    Anti-aircraft fire at the single German plane that flew over Omaha Beach several
nights.**

**Figure 2.15  LCI serving as a hospital ferry.**

**Figure 2.16    Another view of the blockship breakwater at Omaha Beach.**

**Figure 2.17  "Phoenix" breakwater at Omaha Beach.**

We turned our phoenix over to the harbor tugs and returned to The Solent, bringing a few casualties with us.  One of the stories told by a master sergeant who had a minor leg wound was interesting.  He was on a barge loaded with Red Cross trucks which was being ferried to shore on D-Day plus 2.  A few shells were plopping into the sea around the barge but he was not concerned.  There was an Army doctor on board, however, who seemed to be quite perturbed.  He was walking back and forth on the stern of the barge and sweating profusely.  The sergeant decided to cheer him up a bit and called over to tell him that there was no cause for alarm because even if the barge suffered a direct hit they could jump into the water and swim ashore.  The doctor looked at the sergeant for a moment and then said, "Come with me, son, I want to show you something."  He strode to one of the Red Cross trucks and opened the back doors.  The truck was packed with incendiary bombs.  The sergeant said that the doctor then sat down and he (the sergeant) took to pacing the deck.

Back in England I learned that several of the second wave of phoenix tows had been attacked by E-boats out of Le Havre and sunk.  In one or two cases there were no survivors.  It was assumed that the men had been thrown off the anti-aircraft-gun platforms by the torpedo blasts and killed by the fall of 60 feet to the bottoms of the units.  Then they were probably entombed in the units.

On D-Day plus 12 I collected my meager belongings aboard the Queen of Thanet, said my good-byes to my British friends, and boarded another tug for duty in France. Again we had a phoenix in tow. Sticking to the mine-swept channel, we crossed to the Bay of the Seine without incident. We arrived, however, just about the same time that the worst June gale in 40 years hit the beachheads. The colossal damage done by this storm has never been fully reported to the U.S. public. In my opinion we came closer to losing the war in the three and one-half days that the storm lasted than at any other time. If the Mulberry harbors had not been put into full operation a few days earlier, and if it had not been possible to unload the giant LSTs by beaching them under the calm conditions created by the breakwater units from D-Day plus 3 until D-Day plus 13, perhaps we <u>would</u> have lost the war. Many high-ranking officers I talked with later held the theory that the vast amounts of materiel landed from D-Day plus 5 to D-Day plus 13 was the crucial factor in the successful Cotentin Peninsula campaign.

**Fig. 2.18**     **First road cleared of mines through the Omaha Beach bluffs.**

Fig. 2.19      LSTs taking advantage of calm seas created by the Mulberry breakwaters

Fig. 2.20  Full-scale LST unloading activity at Omaha Beach.

Fig. 2.21 LST approaching a "spud" pierhead at Omaha Beach.

Fig. 2.22 LST docking at a "spud" pierhead at Omaha Beach.

**Fig. 2.23 "Spud" pierhead upper ramp.**

**Fig. 2.24 First Mulberry unloading activity at Omaha Beach.**

**Fig. 2.25  Trucks on a "whale" causeway at Omaha Beach.**

**Fig. 2.26      More materiel going ashore at Omaha Beach over a "whale" causeway.**

The storm was so bad that we could not release our tow.  For three days we had to steam through the mine-infested waters northwest of Omaha Beach with a 6000-ton albatross tied to our tail.  The waves were so high we were shipping green water over the bridge.  As we mounted the crest of each huge wave and started down into the succeeding trough I was sure that we were going straight to the bottom.  We were completely out of food but I didn't care one bit.  For the first time in my life I was seasick -- so sick that I even considered swimming the mile or two to shore.  The captain of the tug refused to let me try, however.

Each time we came close to the Mulberry harbor, I roused myself from my giddiness sufficiently to study it closely.  It was almost finished.  Two causeways were connected to pierheads and the phoenix breakwater was 50 units long.

During the first day of the storm the harbor seemed to stand the fierce buffeting rather well.  Then on the second day some of the bombardons tore loose from their moorings and came crashing into the ships of the outer roadstead and into the line of phoenixes.  Some of the battered ships also tore loose and were driven relentlessly into the inner breakwater.  Severely breached, some of the phoenixes began to crack up.  Once they started to crumble it did not take long for the heavy seas to finish them, reducing the product of millions of hours of hard work to rubble in a matter of hours.  Then the heavy seas poured in on the light craft in the inner roadstead, hurling them up on the shore and against the Mulberry causeways.  By the end of the third day the wreckage was indescribable -- piled up for five miles along the beach.  Only pictures can do this holocaust justice.

**Fig. 2.27     Waves of the D-Day and 13 storm hitting the Mulberry breakwaters.**

Fig. 2.28        Ferocity of the D-Day and 13 storm soon began to weaken the Mulberry
                 breakwaters.

Fig. 2.29        Damage done by the D-Day and 13 storm was probably the greatest marine loss
                 in history.

**Fig. 2.30      Many small craft were driven into the Mulberry causeways by the storm.**

Watching it occur from my precarious and wretched position at the rail of a violently pitching tug, I was as low in spirit as a person can get.

But, as always, misery ends and the sun shines again. On D-Day plus 17, a Seabee duck wallowed through the calming seas and took me ashore. When I landed I must have been a pitiful sight. Almost too weak to stand up, I had no weapons, no helmet, no gas mask, nothing to indicate I was anything but a poor refugee except a motley uniform consisting of a khaki-covered cap, a fur-lined blue jacket (remember that this was in the month of June), khaki trousers and shirt, marine boots and a black tie. The only luggage I had was a soft leather valise filled with underclothes and toiletry articles. I shall never forget the greeting I received from the Seabee officer who met me at the water's edge: "Well, the first carpetbagger in France!"

The story of the clean-up after the storm is too long to tell this afternoon. Let me close by telling you briefly what was done with each of the two harbors. The American harbor was patched up to serve as a shelter for small boats; the British harbor, which had not suffered the full force of the storm and had not been as far along in construction as the American harbor had been when the storm struck, was relatively undamaged and was completed sufficiently to serve as a port of entry through the remainder of the French campaign. The American harbor, however -- primarily due to the superhuman efforts of Captain A.D. Clark,

Commander of the Mulberry Task Force, and the men under him who actually built it will live forever in my memory as the macro-project that played a crucial role in the winning of the war against the Nazis.

**Fig. 2.31  King George VI and Admiral Kirk, USN, inspect Omaha Beach.**

# 3

# Financial Engineering of the Channel Tunnel

**André Bénard**
*Chairman, Eurotunnel*

When I was first asked to address the Macro-Engineering Research Group, I was somewhat daunted by the prospect of addressing such an audience on a purely technical subject. On thinking about the complexities of any large engineering project, be it mechanical, civil, or design, it struck me that the common factor running through all types of project is money. Success in any sophisticated endeavor has to depend on the structure and quality of its financial support. The greater the undertaking the greater the dependence on its financing. By putting the emphasis on the financial engineering aspects of the project, I open up the opportunity to give you some of the flavor of the problems we encountered and surmounted during the course of construction.

Eurotunnel is a very large project, some £10 billion or $15 billion. It does not involve new technology, but it brings together many technologies, some of which are quite new, and it could figure in the Guinness Book of Records from a number of points of view.

I could quote many firsts in this project, such as:
- its conception, in the absence of any client;

- the non-existence of legislation on safety issues for this kind of infrastructure;
- the fact that it was 100% privately funded with no recourse to public money or government guarantees;
- it was built by a bi-national company, under the close monitoring of a bi-national intergovernmental commission;
- because of its size, it had to call on a completely international syndicate of some 220 banks.

From a technical point of view:

- it is the largest undersea tunnel built to date;
- the rolling stock has to accommodate three types of electrical current - picked up in the UK from a third rail and, on the continent, from a catenary. It also has to accommodate three different signalling systems - in Belgium, the UK and France;
- the width of the normal TGV carriages has to be reduced to be compatible with the distance between tracks prevalent in the UK;
- it uses the heaviest rolling stock ever built;
- it has unusual aerodynamic problems resulting from the fact that the Channel Tunnel is really three tunnels connected with each other at intervals.

**Fig. 3.1 The Channel Tunnel Train** (photograph reproduced by permission, from *Le Tunnel sous la Manche* by Bertrand Lemoine, *Le Moniteur*, Paris, 1994)

The concept of a Channel Tunnel has been alive for a long time. The first detailed proposal came from a French engineer, Mathieu-Favier, who presented his plans to Napoleon soon after the Treaty of Amiens in 1802. His idea was a tunnel ventilated by chimneys rising above the waves and lit by gas lamps. Then there was the scheme of Thomé de Gamond 150 years ago, who attempted the first geological survey by diving from a boat to take samples - with weights on his feet (pig bladder floats), ears full of fat and fabric and a mouth full of olive oil against water pressure. In the process, he had to fight a number of enemies: the cold, the sea, his contemporaries' skepticism and also some large conger eels who were not enlightened enough to understand the importance of his endeavors.

Since then, there have been several more attempts to progress a fixed crossing of the Channel. One of them led to tunnelling that began on both the British and the French coasts with Colonel Beaumont's machine, a forerunner of today's Tunnel Boring Machines. None of them proceeded very far, principally because the British military establishment considered a tunnel to be a threat to Britain through invasion. It was not until 1955 that the British government announced that a tunnel was no longer a threat to the country's security.

**Fig. 3.2    Col. Beaumont's original tunnel boring machine, 1881** (photograph reproduced by permission, from *Le Tunnel sous la Manche* by Bertrand Lemoine, *Le Moniteur*, Paris, 1994)

**Fig. 3.3 The Robbins boring machine chosen by the French Channel Tunnel Company**
(photograph reproduced by permission, from *The Tunnel: The Story of the Channel Tunnel 1802-1994*
by Donald Hunt, Images Publishing (Malvern) Ltd., U.K., 1994)

Once again, plans and schemes for a tunnel began. They culminated in an
Anglo-French Treaty which allowed preparations for a tunnel to·begin in 1974
backed by private and public finance. However, a year later the British Socialist
government of the time reneged on the agreement on the grounds that they were
unwilling to accept the commitment for the financial guarantees considered
necessary. Neither did they wish to allocate funds for the high- cost railway link
that would be needed between the Tunnel and London. There was the added
doubt that the British Parliament would ratify the Treaty. It was not surprising
that there was considerable annoyance in France and among private investors at
the British decision to withdraw.

With the arrival of the Thatcher government in 1979, the atmosphere changed,
and Britain and France once again did a great deal of preparatory work for putting
into place another Fixed Link Treaty.

In 1981, President Mitterrand asked for ideas and proposals for the provision of a fixed link and, at the same time, initiated studies and reports from independent institutions as to the practicalities and viability of such a project. In 1984, Margaret Thatcher gave her support to the project on the condition that it would be entirely privately financed.

Banks were asked to report on their degree of certainty in the ability of the private sector to "perform", given the many unknown factors associated with such a large bi-national project. For example: tenderers had no experience funding such a venture, there was no dedicated equity base, and the payback period would be unusually long.

The banks reported back in May 1984 - about a year before formal bids were asked for - and in essence said the following things:

First:      present technology was such that the private sector quite probably was capable of completing the construction of a Channel crossing.

Second:     a tunnel was the least risky technologically and therefore was likely to be the most attractive offer in the raising of private finance.

Third:      the major funding would have to be bank loans.

Fourth:     the banks would have to be closely involved.

Fifth:      the governments should be the lenders of last resort.

This last condition, which was not in line with Mrs. Thatcher's brief, was replaced by a lengthening of the duration of the Concession.

The completion and assessment of all the studies culminated on November 30th, 1984 when Britain's Prime Minister and the President of France issued a statement to say that "a fixed cross-Channel link would be in the mutual interests of both countries."

Politically, the timing was right for both France and Britain. Mrs. Thatcher was beginning her second term and wished to emphasize Britain's commitment to Europe when it not always appeared so. Also, by her insistence that no public money or guarantee was to be involved in this project, she wished to state her belief in the capability of the private sector to undertake a major infrastructure - hitherto the sole province of government funding.

France, on the other hand, wished to increase technological progress particularly in areas of high unemployment such as northwest France.

By the beginning of 1985, both governments were ready to ask for formal bids with the condition that any schemes submitted must be originated by a truly bi-national consortia. Formal invitations to tender were sent out in April with a closing date at the end of October. The decision was to be announced by an Anglo-French Committee on January 20th, 1986.

Out of the nine bids submitted, four went through to the short list:
- motorway suspension bridge.
- a tunnel carrying both a road and a railway.
- a submerged tube with a suspension bridge for part of the distance.
- twin rail tunnels, separated by a service tunnel, to carry through trains as well as specially designed shuttles to transport cars and trucks on a closed circuit between the tunnel terminals.

On the due date in January, 1986, the French President and Mrs. Thatcher together announced that the successful tender was the twin rail tunnels submitted by France-Manche and the Channel Tunnel Group. These two companies were specifically formed in order to bid for the concession. They consisted of five French and five British contractors supported by three French and two British banks that were to act as the arranging banks for credit agreements.

The following month, Britain and France signed the Fixed Link Treaty and in March, 1986 the two companies signed the Concession agreement. This gave them the mandate to design, build, and construct for delivery a fully operational system. It also allowed the concessionaires to operate the Channel Link until the year 2042 - with freedom to fix their tariffs on an exclusive basis for the first twenty years.

So, at the outset, the ownership, building and operation of the tunnel was being undertaken by a consortium of fifteen strong-minded and individual companies, five of whom (the banks) owned 40% of the founder shareholders' shares, while the 10 contractors owned 60% of the shares.

Early on, it was recognized that such an ownership arrangement meant that the organization would appoint and instruct itself -- a situation that would lead inevitably to disruptive conflicts of interest. To overcome this, it was clear that it was essential to create an independent structure to take over the responsibilities of the Concession and to manage the project as a whole, leaving the contractors to build the system.

It was at this point in May, 1986 that the founder members created the Eurotunnel Group and seconded the personnel to run it - together with its own Board of Directors. The foundation of Eurotunnel took several months. When it was really created, in September 1986, the main construction contract had been signed. The basis of the loan agreement and of the Railway Usage Contract had been agreed, the first tranche of shares - Equity 1 - had been issued to the founders in the proportions indicated above. The first equity, some £46 million, was in essence the capitalization of the early expenditure incurred by the founders.

At the same time as the arrangements for Equity 1 were being made, work on Equity 2 was being progressed so that it could be launched the following month in October. Equity 2 was a private placement of £206 million and provided for the ongoing costs until such time as the main credit agreements and public issue - Equity 3 - could be arranged through the latter half of 1987.

It is important to understand the situation that existed in the run up to Equity 2:

(a) Although the Concession and the construction agreements had been signed by the founders, the Treaty had not been ratified by either government. Neither had the enabling legislation been passed.

(b No credit agreement was actually in place although 40 banks had confirmed their commitment - in principle only - to underwrite such an agreement. However, the banks made the condition that the commitment was valid only if the major public issue - Equity 3 - took place the following year. In addition, no listing was available for the shares to be subscribed.

(c) No usage contracts had been signed, although terms had been agreed with the British and French national railways and were a condition of the signing of the Agreement. The banks made it conditional that the contract be signed before loans to Eurotunnel could be made. The Railway Usage Contract was of great importance from the financing point of view, in that it reserved 50% of the tunnel's capacity for the two national railways from which Eurotunnel would receive tolls and a guaranteed payment once the tunnel became operational.

An additional local difficulty was that the banks did not agree with the *Maître d'Oeuvre* and appointed an independent consultant.

By the time I joined Eurotunnel as Co-Chairman in September 1986, the contractual arrangements mentioned above had already been drawn up. So, as much as Eurotunnel would have liked to re-negotiate many items - or, indeed, to start again - it was not possible under the terms of the Treaty or of the Concession. Furthermore the funds contributed by the shareholders were all but finished. We could not but go forward with Equity 2 and had to be satisfied with such amendments as we could achieve without alterations to the fundamental principles of the project.

With all the potential risks in the areas of politics, finance and operation it is not surprising that it needed considerable powers of persuasion at all levels to complete the placement of Equity 2, which, if it had failed, would have brought the whole enterprise to collapse.

It is worth remembering who negotiated such agreements as were in place at the time of Equity 2. They were the founders themselves, who were the interested parties in the construction contract, and the banks, who were interested in the Credit Agreement and the embryo Eurotunnel. So up to the placement, the founders effectively controlled Eurotunnel with seconded specialist staff who, by the nature of their employment, might well have a conflict of loyalties when acting for new independent owners.

Banks generally need the comfort of constant reassurance that there is an owner who has a strong cash flow behind any project that they support, which at that time Eurotunnel did not have. I believe that banks involved in other project financings, having noted our experience, are now seeing to it that there is a proper owner with the right authority in place from the start.

Had we, as Eurotunnel, really been able to renegotiate, I am sure that a great many of the disagreements that later occurred between us and our contractor, TML, could have been avoided. As it is, we look back from 1993 over a period which has been characterized by bitter and prolonged disputes between contractor and client, most of them stemming from arguments over the lump sum part of the design-and-build contract, as a result of which the specifications had to be in terms of performance rather than design.

With the benefit of hindsight, I think it probable that the disputes were no more than the inevitable consequence of a project that was launched before it was ready. It would have been better to have spent the necessary time on getting the design details right, including the safety requirements of the Intergovernmental Commission, and getting the budget right. As it was, the financial pressures demanded to have the project in operation within a little over seven years from the award of the 1986 Concession.

The lessons learned from this are two-fold: first, that more time spent in the early design stages might well have saved us a number of the contractual problems that so bedeviled the last few years; second, that the time scale and provision of the initial funding is itself a major issue whose importance cannot be over-emphasized.

In the first part of 1987, considerable advances were made toward securing our future. In July, the Parliamentary processes in Britain and France were completed so with the Treaty ratified the political risk was removed. Also in July the Railway Usage Contract was signed and, with the financial risk reduced, in August and September 50 banks underwrote a £5 billion Credit Agreement and the European Investment Bank signed a further £1 billion credit - the largest to which they have so far subscribed.

However, before the way was clear to proceed with Equity 3, a further stumbling block had to be removed. The French banks were adamant that they were unwilling to support a loan agreement unless there was comfort that the high-speed train system was extended to the French tunnel entrance at Calais. Without such access for cross-Channel traffic into the proposed French high-speed network, they did not believe that Eurotunnel could operate at its proper efficiency and viability. Clearly, without the support of these banks the loan syndication would not be feasible. It took a year for the French government to be persuaded that, notwithstanding what had been said before, its decision on the TGV Nord and its extension to Calais was of vital importance to the creation of a European high-speed rail network and that it would open the door for expanding abroad French TGV technology.

The French government's decision eventually came three days before contracts were signed with the 200 banks who took part in the syndication which, added to FF10 billion of equity, was intended to cover all the costs of the project until breakeven. The EIB provided FF10 billion of treasury covered by letters of credit supplied by the syndicate. With hindsight, one may reflect on the wisdom of bringing together such a large and complex syndicate whose decisions are taken by 65%, 90% or 100% majority votes. Had the costs been in line with budgets, there would have been no problems, but the very large overruns which we had to live with eventually made the exercise a kind of nightmare. As it was, this is what we got as a result of seeing 550 banks on all continents. With the syndication in place, the way was clear to go ahead with the public offering of the £770 million for Equity 3.

The actual launch of the offer, consisting of indivisibly twined French and English shares, was made on November 16th, 1987 - hardly a propitious moment, being three weeks after Black Monday. On the other hand, we had cause to be encouraged. The underwriting had been achieved in that difficult climate and the fact that over 80% of the issue was taken up by investors indicated that, in our home markets, the sentiment and perception of the project had radically improved. There was a certain amount of astonishment among the public that we proceeded to the offer so soon after Black Monday. Actually, the decision to do so was forced upon us because the funds position was critical. As it turned out, the share price regained its value after three months and the underwriters had no reason to regret the risk they had taken.

All in all - and taking into account the uniqueness and the dual nationality of the project - we view the financing and the progress made in the comparatively short time of two years as a great achievement. To me, one of the principal features of this success is the international spread of the funding - both the equity and the debt.

During the early part of 1988, the construction work got fully under way. Unfortunately, progress did not come up to expectations, partly due to technical problems. There were also organizational difficulties in both the project supervision and in TML.

In fact, by April the situation was such that Eurotunnel made official representations to TML concerning the delays in the tunnelling operations and the slow start of the studies concerning the transportation system. The project expenditure had risen to around £700 million by November 1988 requiring the first drawdown under the main Credit Agreement. It was at this point that the banks insisted that organizational changes be made before they would sanction the next drawdown which was due two months later.

In the searching re-appraisal of management that took place in the autumn of 1988, there were many managerial casualties and resignations in both Eurotunnel and TML, which severely strained the day-to-day running of the organization and the supervision of construction operations. However, the end result was satisfactory in that the rates of progress improved and the program - although not back on schedule - only slipped one month, to a June rather than a May 1993 completion.

Nevertheless, costs escalated quickly, and although the problems of meeting the technical conditions required by the banks seemed containable at the end of 1988, by July 1989 it was clear that Eurotunnel would not have sufficient funds to complete the project. In fact, by the end of 1989 the estimated cost of constructing, fitting out, and financing the project had escalated from the projected £4.6 billion to approximately £7.4 billion - about a 40% increase.

So from July to November 1989 we operated on the balance of our equity and on monies advanced by the banks under a monthly waiver of the conditions of the loan agreements. By this stage of construction, we were using about £100 million a month.

With such a large increase in likely costs, we negotiated with TML for some time in order to resolve the financial difficulties. By early February 1990, we had reached Heads of Agreement for both a change in the management structure and that TML would bear 30% of all tunnelling costs above new target levels. By this time it was vital that the deal should be finalized and signed as soon as possible because the whole enterprise was again in imminent danger of coming to a complete stop through lack of funds. At this point, the banks insisted that the Heads of Agreement should be incorporated into a full contract amendment and set the date of February 19th, 1990 for signature. February 19th came and went with TML reneging on the Heads agreed less than a month before. So I arranged a board meeting in Paris for the morning of February 20th and put liquidation on the agenda. It was not until 5 o'clock that evening that TML contacted my colleague, Alastair Morton, and myself, thought fit to check the views of individual Eurotunnel directors, and after some further discussions the amendment was signed at 9 o'clock that night.

I think it is of interest that TML's exceptional concession to share part of the expense of tunnelling overruns was the key to the banks' agreeing to continue.

Without it, they believed that they would be bearing an unacceptable proportion of the risk.

Once the amendment was out of the way, the banks extended their waivers while Eurotunnel set about raising the further funds that were now considered necessary to complete the project and commence operations. Their task was not made easier by the fact that Eurotunnel, TML, and the advisers appointed by the banks could not agree on how much the costs would overrun. Furthermore, the considerable claims that TML was making against Eurotunnel had yet to be substantiated and agreed.

It was finally settled that the total cost of the project to completion meant that a further £2.7 billion was needed. This was to be provided by syndicating a £2.1 billion loan and raising a further £560 million through a rights issue. However, within this total figure was £1 billion of funding margin which was considered necessary, both to reassure investors and to satisfy the bankers' cover ratio. Finding this order of money in the financial climate that prevailed in 1990 was an uphill struggle, added to which Eurotunnel asked that the date for the maturity of the loans be extended from 2005 to 2010.

By late April 1990, plans had advanced sufficiently for a letter to be sent to the shareholders explaining the need for a rights issue. Then followed a period of immensely hard work which one month later resulted in Eurotunnel's agreement or negotiation for:
-    the underwriting of the Rights;
-    a £300 million loan, a so-called parallel line, from the European Investment Bank not covered by the syndicate;
-    the Bank Information Memorandum requesting the additional funds from the Banking Syndicate on a 21-year term basis;
-    a new bank waiver pending syndication.

Between October and December, each of these tasks was completed. Considering the investment climate in which the Rights issue took place, I think it was a remarkable success that it was subscribed to 97%. We ended with 630,000 shareholders - 510,000 in Paris and 120,000 in London. With respect to the loan, the fall of the Tokyo market considerably restricted the Japanese banks in their lending. Indeed, the Japanese banks needed some persuasion to subscribe their share of the new financing. Although there were high-level expressions of political support for the project from the governments of Britain and France to their Japanese counterparts, it was a tribute to the Japanese part of the banking syndicate that they came forward and continued to represent nearly one-quarter of the total financing of Eurotunnel. This was in spite of the fact that overseas loans by Japanese banks had dropped by one-third in 1990 compared to 1989.

After this major refinancing, the progress of the tunnelling remained good and June 1991 saw the breakthrough of the main running tunnels.

At last, there existed a fixed route between France and Britain. It had taken 240 years since the idea was first suggested - and 190 years and 27 attempts from the time of the first detailed plans to the achievement of joining Britain to the Continent.

Sadly, by the end of the year, TML experienced difficulty maintaining its progress at the existing rate due to an allegedly increasing shortage of its own funds. Therefore, once again Eurotunnel and TML began negotiations and started the arbitration process foreseen by the contract in an endeavor to solve the problem. In March 1992, a pre-arbitration panel awarded TML £50 million equivalent a month from Eurotunnel on account.

This award was quashed by arbitration in September 1992, by which time £200 million had been advanced. TML retained this money against outstanding claims, but the claims issue proved impossible to settle by negotiation. The commissioning process suffered serious delays. The opening date became unforeseeable.

It became clear that because of the delays, the date when we expected to receive a revenue flow from the start of operations would be badly missed. The ensuing effect meant that we were unable to meet the projected schedule of interest and loan repayments. Eventually, at the end of July 1993, a protocol was signed between Eurotunnel and TML. It provided for a loan of £200 million to be made by Eurotunnel to TML, full cooperation between the parties on target opening days, a handover of the works from TML to Eurotunnel on December 10th, 1993 to execute the tests on completion while TML remained responsible for the overall performance of the system.

Since the signature of the protocol, things went well, but the consequence of the delays was that Eurotunnel needed further funding to reach breakeven point after opening. So in October 1992, we announced that further monies would be needed after the opening, and in April 1993, we followed this by saying that the expected new date for the beginning of freight operations would be in March 1994 to be followed in May by passenger shuttles and later in the year by through trains operated by the national railways direct from London to Paris. Additional research into the traffic volumes and income that we could expect was completed and also the consequent effect this would have on our projected cash flow.

In October 1993, we announced that extra funding of "under or around £1 billion" would be required to carry this project beyond the 1994 completion, and discussions with the banks on the resolution of that final financial problem continued for some time before reaching resolution.

I have spoken at some length on the various stages of financing we went through in bringing the Eurotunnel transport system to its final stages. No one expected that the financing - or indeed the engineering - of such a unique concept would be easy, and as you have heard it has not been.

Now I would like to expand on the lessons learned from our experiences, in the hope that future privately financed infrastructure projects may perhaps avoid some of the anguish that we experienced along the way.

First, I will expand on the question of ownership. You have heard that in the first instance there was no client. The contractors were promoter, contractor, and client, and had they remained so, they would not have received the support of the banking fraternity. The arranging banks did not like blurring the roles of client and contractor, and they said that had such an arrangement remained, it would not have been acceptable to potential subscribers to any syndicated loan. Consequently, the banks stipulated that an independent client should be put in place.

For the embryo Eurotunnel, coming late on the scene, there were massive problems to overcome. To name just some: there was the recruitment of sufficiently capable men to take forward a macro project of uncertain future; the task of raising equity; the negotiations leading to internationally syndicated very large credits; and the managing of the political scene in the run up to the ratification of the Treaty.

A further constraint on Eurotunnel's freedom of action was that the banks, in a prior agreement with the contractors, had stipulated that all contracts should be placed with or through TML. TML was comprised of civil engineers and therefore the freedom of Eurotunnel to engage other specialists, such as rolling stock manufacturers, was severely restricted, as well as expensive because the contractors were paid on a cost-plus-fee basis and consequently benefitted from price increases.

In listing the principal problems, I have endeavored to explain that in major projects it is essential that there be a substantial owner or promoter who, in their own right, can give credibility and support to a proposal.

There are two types of organizations in the private sector that can provide such assurance. The "stand-alone" owner is one, and the other is a major established -- perhaps multinational -- corporation. The arguments in favor of the stand-alone owners are readily identifiable:

- It clearly brings the sharpest focus on the management of the project.
- It invites the highest degree of public interest.
- It provides the opportunity for equity investment to be directly linked to the success of the project.

- For all these reasons, it brings to bear the maximum pressure to perform both on the owner and on the contractor.

There is little argument as to the desirability of each of these characteristics. On the other hand, there are potential advantages of launching the project under the auspices of a major established corporation:

- It is easier to find competent entrepreneurial management needed for the initial stages.
- Disputes can be handled in a more serene environment without the glare of publicity.
- It provides treasury support if needed during the initial stages, and may indeed provide the "seed corn" itself.
- It provides a financially muscular guarantor of the owner's obligations, and thereby reduces the risk of the owner being backed up against the wall of financial collapse by contractors seeking to extract concessions.

In the case of Eurotunnel, we were a stand-alone owner albeit handicapped at the outset.

Our existence covered two different phases in the evolution of the project. The first was the construction period, marked by high levels of risk as to technical feasibility, construction cost, timetable, etc.; the second was the commercial viability of the ultimate system.

In the construction phase, secured lenders were looking for a high degree of assurance as to the amount of the cushion that lay between their exposure and the economic collapse of the enterprise. In theory, that cushion can be represented only by equity or prohibitively expensive subordinated debt. In the absence of such a cushion, senior debt came at a high cost, which of itself places a heavy burden on the project's economics.

When the initial operating troubles are overcome, and the project settles into its ultimate destiny of being a very large transport system, with easily determinable costs and a highly reliable revenue stream, it can reasonably expect to approach the financial markets for revolving debt at attractive rates of interest. Between these two phases, there will clearly be an interregnum while the high-risk construction project is emerging into the low-risk system, and that period also requires recognition of its own needs and constraints.

Eurotunnel was constructed with a straightforward balance sheet initially consisting of £4 billion of syndicated debt plus a £1 billion standby and £1 billion of equity. There was no cushion for the banks. Near the end, they were looking for a high level of equity finance both to alleviate their risk and to reduce the interest burden. The nature of ownership has, therefore, a direct influence on the cost of the project.

I would now like to expand on another vital ingredient in the relationship of owner to contractor, that is, the contractor's duty to the organization to which they are obligated. Eurotunnel's position at the outset was complicated in that the contractors set up Eurotunnel having already signed the construction contract with themselves, as it were. Having installed Eurotunnel as owner, the contractors then demonstrated a cavalier attitude toward the quotations which they had produced themselves. They obviously did not feel bound by their own estimates.

As a result, we had to return to the market in 1990 for a 40% increase in our funding, and again became mired in negotiation for the extra funding needed to carry the project from completion to breakeven.

Clearly, private funding cannot easily accommodate the very large overruns, both in absolute and relative terms, that repeatedly occurred. Public finance may be more accommodating, but the importance of properly established budgets is vitally important. They should be the first and foremost preoccupation of the authors of a project or their contractors. That was obviously not the case for Eurotunnel.

We wrote to the contractors in early 1988 stating that it was our opinion that the complexity and time scale for the building of the transport system had been seriously underestimated. From the outset the banks had foreseen that aspect of the problem. That is the reason why the fixed equipment had been treated as lump sum in the construction contract. The parameters within which the price could be revised had been clearly defined.

In the case of Eurotunnel, our contractors asked that the contractual lump sum agreement be replaced by the payment of a "global" sum, on the grounds that the justification of each of their claims was too complex to be contemplated.

Can fixed price bids for such projects be trusted to protect the client? What amount is wise for the client to provide for the availability of stand-by facilities to lie beside the principal funding? Will the provision of large stand-bys result in higher costs because contractors know of their existence?

When one has lived through the Channel Tunnel battles, one cannot help but ask such questions. To this day, I cannot say that I have totally satisfactory answers to them.

The following may provide some clues:

First:    Governments that, through safety or other regulations, have a great deal of responsibility for the overall cost of a project, must be made accountable for their decisions.

Second: In planning major works perhaps it might be possible to assess contractors on such things as past performance for the reliability of their estimates and their record of completing on schedule. From this a coefficient could be calculated which could be applied to bids submitted by various contractors.

Third: It may be that in planning major infrastructure projects the Tokyo Bay project might be used as an example. In commissioning the Tokyo Bay Bridge, the Tokyo Highways Authority accepted a bid in which the construction contractors also became the operators and the agents for the Highways Authority. In this way they avoided the conflicts of interest between builders and operators, but did they safeguard the interests of the investors?

It must never be forgotten by the promoter of a major infrastructure enterprise that his money is made from the long-term success of the project, while his contractor makes his money from the short-term construction period. When the contractor is both contractor and operator, to which role will he give precedence?

In our case, our contractors were given low price shares and very generous warrants in the belief that such goodwill would encourage them to make sure that they would overcome such difficulties as they might have in a way that would not interfere with their interests as shareholders. This was a miscalculation because of the very point I have just made: that contractors looked first at their construction contract, and it soon became apparent that their actual shareholding was of secondary importance.

Turning now to one other significant lesson that has come out of our venture: if you can avoid it, do not have too many banks involved in a major project. We had 233 banks spread across almost every country in the world that had any sort of banking tradition. The size of the Eurotunnel syndicate occurred as a consequence of the determination of the 50 underwriting banks to finish up between them with no more than 50% of the ultimate commitment. To achieve this, they had to bring in another 173 banks with widely differing degrees of sophistication in project finance matters.

Accordingly, we found ourselves with the last 5% of our facility being spread among no fewer than 65 banks. Many of these banks had little, if any, experience with complex project financing, and frequently had great difficulty producing enough people with the necessary experience to work through the mass of paper that flowed across their desks.

Our syndicate was led by the four Agent Banks - NatWest and Midland in the UK and Credit Lyonnais and Banque Nationale de Paris in France - who established a permanent team to monitor the project and maintain communications with syndicate banks. We in Eurotunnel also established a bank relations group, through which we aimed to meet every one of our syndicate banks at least once a year, and to maintain regular contact with the larger of them.

The communication problems posed by a syndicate of this size were immeasurably compounded by the rigid structure of the Credit Agreement itself. The documentation ran to some 1,500 pages. It called for compliance with over 50 warranties and over 200 covenants on the occasion of each drawdown from the facility. It imposed strict constraints on the extent to which we could draw down between the five tranches into which the facility was divided. It involved a vast edifice of successive layers of supervision and scrutiny, and left the leading Agent Banks with very little discretion to exercise judgment on the application of multitudinous and restrictive provisions. Most important decisions had to be approved by at least 90% of the syndicate by value, and in a number of cases by an almost effectively unachievable 100%.

At the same time, in the years since it was put into place, the world has changed dramatically. In particular, the world of banking has changed, with the result that facets of the facility that were seen as being flexible and forward-looking in 1987 have become completely unworkable in a world in which few banks are prepared to carry the risks of supporting other banks' obligations.

None of this is to detract from the immense support that we in Eurotunnel received from our banking syndicate, nor to under-estimate the difficulties that many of them found themselves in during a time in which banks' own balance sheets came under unprecedented strain.

I make the point only to press for recognition of the fact that circumstances change over the long periods covered by major project credits. In the ideal world, it is best to have the smallest possible syndicate, made up of banks with real competence in project finance matters, that can provide the Agent Banks with a useful degree of discretion to make business decisions.

I have explained in some detail the complexities of financing that have to be considered in these days of macro-engineering projects. The last "lesson learned" about too many banks brings me to my final observation about the future financing of the infrastructure. There is no question but that the private sector is fully competent to carry out such works that previously have been the sole responsibility of governments.

It is over 200 years since Adam Smith, the Scottish philosopher and strong advocate of private enterprise, examined the role of private sector finance for infrastructure projects. He argued that there was no reason for such projects to be "defrayed from the public revenue" provided that they are made "only where commerce requires them and that their expense, grandeur and magnificence must be suited to what commerce can afford to pay." His philosophy certainly was followed in Britain where during his time and in the 1800s there were built 1000 Turnpike trusts and 15,500 miles of railway.

The difference between the 18th and 19th centuries and today is the ever-increasing sophistication and aspirations of the population and the consequent desire for faster, safe communication and larger more efficient utilities.

As those pressures of society's demands continue to build, governments everywhere are finding it increasingly difficult to respond adequately to provide clearly needed infrastructure. They have far greater control over investment today than was dreamed of in the times of Adam Smith but, in broad terms, I believe his proposition to be correct in the sentiment "provided commerce can pay." In other words, provided a project will make a satisfactory return on investment.

With infrastructure, governments have considerable influence in negotiating and supervising contracts; consequently, by their intervention, they assume a significant degree of responsibility - an opinion which, by the way, is held by many banks. This being so, I ask whether it is realistic that governments refuse any form of financial responsibility in an infrastructure project that can be shown to give a satisfactory return on investment in its own right.

For some time our proposition has been that, under such conditions, there is a need for a form of "mezzanine" finance to stand between the equity and the senior debt. This could be provided by government guarantee, which would only be called when all loans and stand-by credits had been exhausted.

Such a guarantee would have several advantages:

- the government would be involved from the start in assessing the degree of return that can be expected;
- the guarantee need not be subscribed but its existence would allow the government a commercial interest and therefore the power of intervention;
- taxpayers' money need not be disbursed, thus allowing the government's credits and funds to be used elsewhere;
- the presence of the government guarantee would ensure the credibility of the project in its funding. But this kind of government role should not impinge on the freedom of the operator to fix his tariffs according to market conditions.

In summary, it is our opinion that the state could do a great deal better by providing this form of help than by directly investing large sums of money for no perceived commercial return. From the governments' point of view, they would not only be the probable sponsors for the scheme, they would also be part of the process of assessing the commercial return. Provided this was satisfactory, the guarantee is likely to be a low risk, whereas the political and economic benefits would be very large.

Such an approach would clearly establish the reality behind a privately financed infrastructure. It is in fact a partnership between governments, contractors, banks and investors. In such a partnership every partner has to play his role.

When hard times appear, no one can walk away, each party has to play its part in the situation. In Eurotunnel's case, banks and investors played theirs. Contractors solved technical problems but were also the source of the financial ones.

But the main beneficiaries, governments, have originally contributed by granting the Concession. They have not so far helped in solving the problems they have themselves caused by their regulatory demands. It should be part of the original understanding that they do so.

The completion of a Europe-wide High Speed Rail Network within the European single market is an ambitious and important program. The opening of the Channel Tunnel is an essential part of this whole process of integrating the 12 nation states. I say it is an essential part because, if a map is drawn based on high-speed train time taken between city centers, Britain's 60 million people are moved dramatically nearer the center. As 60% of Britain's trade is with Europe, the significance for the UK of the Eurotunnel project is understandable. More important still, at a time when Europe's political future is unclear, the Tunnel and the European high-speed rail network are a concrete and indestructible element in the bringing together of the peoples of Europe.

The last seven years have been a long and sometimes hazardous route. I am often asked: "Would you do it again?" My answer is: "If I had the choice, I would do it differently, but if I had no choice, as was the case in 1986, yes, I would do it again. The challenge is worth the pain and the effort." And it was all good fun.

# 4

# Old Cities and New Towns For Tomorrow's Infrastructure

**Andrew C. Lemer**
*Director, Urban Research Board, National Research Council, Washington, DC*

## 4.1 INTRODUCTION

More than 150 years ago (in 1841), construction of the Thames Tunnel near the City of London was completed. This project, a road tunnel later converted to railway and still serving London's subway, was one of the earliest large-scale applications of shield tunneling. Its completion was later described as a "shining and inspiring example of human perseverance and capability," a work that placed its creator, Marc Isambard Brunel -- civil engineer, mechanical engineer, and architect -- "among the most eminent men of his profession for all times."[1]

The Thames Tunnel is but one of an extraordinary collection of construction accomplishments that mark the 19th century as an "Age of Infrastructure" (although the term "infrastructure" would not be created until nearly a century later). Modern water supply was born in London in the middle of the century and, in 1857, Brooklyn (New York) built the first modern urban sewage treatment system.[2] The first concrete roads followed within two decades the 1824

invention of Portland cement, and the Place de la Concorde in Paris was paved
with asphalt as early as 1835.[3]  Alexander Graham Bell invented the telephone
in 1876, and Edison the electric light in 1880.

This "age of infrastructure" has yielded a rich legacy, supporting a century of
unprecedented economic advance, but today we risk squandering this legacy
through neglect and mismanagement.  In cities around the word, water supplies
have grown less reliable, wastes mount in prodigious volume, traffic approaches
gridlock, and the air grows unbreathable.  "NIMBY" (Not In My Back Yard!) has
become an almost universal symbol of public response to infrastructure facilities
planning, and governments at all levels find themselves hard-pressed to allocate
adequate resources to maintain their current public capital, much less build for the
future.

We are on the verge of a quiet infrastructure crisis of global proportions.  It
is quiet because we fail to recognize, as professionals and public, either
infrastructure's essential importance or the opportunities for enhanced services
that new technologies offer.  It is a crisis of lost confidence in technological
advances as well as outright neglect in a society rapidly losing sight of the public
well-being that springs from public works.  And it is global because all the
world's nations share common needs for new public works infrastructure
technology.

The world's population is fast approaching the point at which more than half
of all people live in urban areas (the figure already averages more than two-thirds
in the more developed nations and exceeds 75 percent in the United States).  If
we hope to provide the crucial enabling environment for economic growth and a
high quality of life, we must overcome this quiet crisis and enter a new "age of
infrastructure."  The lessons of both past and present will help us build the
infrastructure of tomorrow.

## 4.2 WHAT IS INFRASTRUCTURE AND WHY
## DOES IT MATTER?

The term "infrastructure" was coined, according to dictionaries, in the first half
of the 20th century to refer to military installations, and some researchers trace
its origin to Winston Churchill.  Now defined as an "underlying foundation or
basic framework," infrastructure has come to connote a diverse collection of
constructed facilities and associated services,

...both specific functional modes--highways, streets, roads, and bridges;
mass transit; airports and airways; water supply and water resources;
wastewater management; solid-waste treatment and disposal; electric

power generation and transmission; telecommunications; and hazardous waste management--and the combined system these modal elements comprise.[4]

Many of the facilities are built and operated by governments, and thus fall easily into the category of public works; others are built or operated, in whole or in part, by private enterprise or in joint public-private partnership.

Most of the importance of infrastructure is derived from the role of the individual modes in supporting broader social and economic activities. The synergistic effect of infrastructure and social systems is most easily seen in the world's cities; indeed, cities are the ultimate expression of infrastructure and would not function without its services.

The beginning of Rome's water supply by Appius Claudius in 312 B.C. was an early step in development of an infrastructure which by the first century A.D. included paved roads, fire protection, and sewerage that supported Rome's centuries of undisputed dominance. The sewer system of Paris contributed decisively to the decline in severity of water-borne epidemic diseases and became a noted tourist attraction.[5] Advancing transportation technologies enabled the growth of Birmingham, Pittsburgh, and the other great industrial cities in Europe and North America. In the current information age, electronic communications have become an important element of infrastructure, and cities from New York to Osaka seek to support growth by constructing "teleports" linking their businesses to the global satellite network.

The debilitating impact of inadequate infrastructure is notable as well. Production costs for goods and services are estimated to be as much as 30 percent higher in Lagos, Nigeria, and other cities of the developing world, because firms must provide their own water and power supplies.[6] High rates of morbidity and mortality, particularly among the young, are the endemic and costly results of poor management of waste, water supplies, and roads in less developed nations.

While such management continues to generate debate among economists, some analysts attribute a major share of the decline in U.S. productivity since the 1960s to declining rates of public investment in infrastructure.[7]

Neglect of public works investment occurs, in large measure, because infrastructure is hardly noticed until it fails to perform. However, the facilities of infrastructure, linked in functional systems distributed throughout an urban area, typically account for 15 to 20 percent of the land used in a city or town. In a country such as the United States, with its extensive highway network, the proportion can be much larger, averaging 36 percent in eleven major cities.[8]

The total investment in infrastructure facilities accumulates in older cities over many years and is difficult to value, but the planners of new towns must anticipate these costs. For example, in Abuja, the new capital city of Nigeria, with target population of 1.6 million people planned as the world's largest free-standing new town, the investment in local and large-scale regional systems was estimated to account for some 37 to 46 percent of the total investment required to build the plan.[9]   In contrast, infra-structure's share of the total estimated investment in Batam Centre, a modern new commercial town in Indonesia with a population of 147,000 and a central business district, is only 20 percent.[10]   In both cases, technologies specified in the planning were chosen to be appropriate to the levels of technical skills and maintenance capability likely to be available in each city.

Fig. 4.1        The infrastructure of Batam Centre, a new town in Indonesia, was planned to support both global business and local cultural values, while protecting the qualities of a delicate tropical site.

Studies done in the early 1970s for the U.S. Council for Environmental Quality estimated the per capita costs of infrastructure investment in typical U.S. mixed density communities to be $1,500 to $2,000.[11] The 1992 figure is probably closer to $4,500 per capita, on average, and substantially higher in older, denser urban areas. Nationally, our infrastructure may represent a total investment exceeding $1.4 trillion.

The scope of infrastructure changes from time to time. For example, for several centuries all the refuse of Paris, including human wastes, was brought by private individuals to dumps scattered around and throughout the city. These dumps sometimes achieved such heights that during the reign of Louis XIII (1610-1643) they had to be incorporated within the city fortifications, for fear that enemies would use them for gun emplacements during a siege. Growing concern over the dangerous effects of miasmas arising from decaying matter motivated a 1758 royal ordinance requiring future dumps to be located outside the city and, in 1781, Montfaucon was designated Paris' sole dump. Solid waste disposal had become a fully regulated municipal function, an element of infrastructure.

Operating and maintenance procedures, management practices, and development policies, i.e., the "software", are also essential elements of infrastructure. This software interacts with societal demand and facilities "hardware" to determine the performance of infrastructure systems. The success of such corporations as Federal Express and DHL in air parcel delivery, for example, reflects the development of new infrastructure software, producing new services from existing infrastructures and enhanced performance for users.

Besides the water, waste, and transportation systems that comprise the hard core of infrastructure, today's definition may include systems of public buildings - schools, health care facilities, government offices, and the like. These facilities, not as individual structures but tied together by the functional and administrative systems they house, provide important services to the public at large, in much the same fashion as highways and water supply networks.

## 4.3 PROBLEMS OF TODAY'S URBAN INFRASTRUCTURE

In the United States, a decade of studies of the results of neglect have concluded that our infrastructure, if not in ruins, is at least badly frayed.[12] Here, as in most other countries, a myriad of local, regional, and national government agencies, quasi-governmental institutions, and private firms are involved in the planning, creation, operation and regulation of physical infrastructure. The facilities themselves are engineered and constructed with many common principles and procedures, but each of the major infrastructure modes is represented by an extensive and specialized body of technical knowledge, professional and managerial people, agencies, researchers, and informal relationships among these various parties. This institutional complexity inhibits both coordinated action and discussion of the cross-cutting issues of infrastructure and its technological advancement.

Nevertheless, while diverse, the facilities, services, and institutions that comprise infrastructure exhibit significant common characteristics and similar problems. For example, demand for infrastructure is derived from the support it provides for other social and economic activities and, generally speaking, greater numbers of people and higher levels of economic activity means greater demand for infrastructure. However, as rush-hour highway commuters and airport users frequently observe first-hand, performance and capacity of infrastructures are acutely sensitive to the patterns in time and space as well as overall magnitude of underlying demand. The number of people who experience severe congestion and delays during a peak period could be easily accommodated with high-quality service if their travel were more evenly distributed throughout the day. The engineering design of an airport passenger terminal, for example, is typically based on peak demand levels that are 50 to 150 percent greater than would be needed if demand were spread evenly over time, and the factor can be even larger in other elements of infrastructure.

Infrastructure is generally capital-intensive. Because of high initial costs, the commissioning of a new dam, treatment plant, or highway is often a newsworthy event that attracts public attention. The costs of regular maintenance and operations seem small compared to construction but may, over the course of a facility's service life, total much more than the facility's initial costs. Infrastructure managers and elected officials, faced with the challenge of balancing competing public priorities and limited fiscal resources, often find it easy to defer maintenance spending and neglect infrastructure's upkeep. Unfortunately, deferrals speed deterioration and failures of the infrastructure. In sub-Saharan Africa, for example, the problem has reached extreme levels. The World Bank estimated that the backlog of neglected maintenance for roads alone exceeds $5 billion, more than seven times the annual spending needed to keep the roads in good shape.[13]

Nevertheless, infrastructures are expected to be long-lived and are routinely designed to meet demands projected for three decades or more into the future. Most dams, bridges, highways, and other infrastructures endure much longer. For example the Brooklyn Bridge is still performing well after more than 100 years, and the Alicante Dam has survived nearly four centuries. (Of course, long life is not assured: the Grand Teton Dam failed immediately when the reservoir was filled, and the Tacoma Narrows Bridge, famous for its dynamic response to winds, collapsed four months after its opening!)

Reflecting the expectations of their planners, U.S. highway pavements are typically designed for 20-year lives. Many fail sooner because traffic loads grow faster than anticipated. In contrast, the German autobahns intended to serve Hitler's 1000-year Reich are, for the most part, still serviceable 50 years later.

The technologies of infrastructure often outlive the facilities. The evidence shows that there is a relatively long time period - on the order of 100 years in the case of rail and road - in the transition from one infrastructure technology to the next.[14] The overall character of today's water supply and sewerage systems would be recognizable to an engineer of the 19th century, although the chemicals and controls used in processing have evolved substantially.

Energy production may be an exception. Power plants have typically been expected to last 25 to 30 years because history has shown them to become noncompetitive by then, although that perception is said to be changing - and lifetimes lengthening - as designers reach the ceiling of thermodynamic efficiency in conventional generation technology.[15] However, new technologies offering higher efficiencies (e.g., gas turbine or integrated gasification combined cycle processes[16]) could spur earlier retirement of these conventional plants. In telecommunications as well, obsolescence currently is more likely than wear or other deterioration to motivate replacements of equipment.

Most infrastructures are linked in networks. Roads and interchanges; water treatment plants, supply mains, and distributors; generating plants, transmission lines, and step-down transformers; sewers, treatment plants, and outfalls: all are tied tightly to one another and to thousands of individual households and businesses. These networks stretch over large areas, quickly transmit changes from one part of the system to another, and the functions of the whole surpass the sum of the parts. Thus, when one transmission line crossing the Potomac River failed one afternoon early in 1992, downtown Washington, DC, was plunged into total darkness. One minor accident on an urban highway can cause miles-long traffic jams during rush hours.

Because of large facility size and network extent, infrastructure often has broad environmental and social impacts, but these impacts have been frequently underestimated or neglected in system planning and management. For example, congested roads in 39 U.S. cities are estimated to have cost drivers more than $34 billion in 1988, in delays, wasted fuel, and higher insurance premiums.[17] In another study, air pollution from motor vehicles was found to be responsible for $40-50 billion annual health-care expenditures and as many as 120,000 unnecessary or premature deaths.[18] Such costs, seldom considered by agencies deciding whether to invest in highways or transit, add perhaps $0.70 per mile to the costs paid by individuals choosing to travel by private auto.[19]

**Fig. 4.2 The designs of the Thomas Road Overpass on a Phoenix, Arizona Interstate was tailored to local conditions, cost less than standard designs, and became a valued element of its community.**

## 4.4 TRENDS FOR TOMORROW

These common problems of infrastructure are unlikely to change for the foreseeable future, but new ones may be added as the result of a number of global trends.

For example, many scientists agree that the next several decades will be marked by a small but very significant rise in the earth's average temperature. Whether this trend is part of a normal climatic cycle or a longer-term response to human activity is still hotly debated, but the consequence in either case will be a shifting both of demands for water and power and of the nature of our forestry resources. These shifts, according to one analysis, could cause water supplies in the Rio Grande valley to drop by more than 75 percent, and the State of Missouri to develop water shortages.[20] New investments and possibly new technologies will be required to respond to these shifts.

Population growth is a similarly important trend, although the primary impact may be felt outside of the United States. United Nations projections indicate that the number of people living in urban areas worldwide could be 4.9 billion by the year 2025, nearly three times the number in 1980.[21] The number of cities with populations exceeding one million, perhaps 250 today, is projected to exceed 600 by 2025. Nearly all of the growth is expected to occur in developing countries, and only four of the world's 20 largest metropolitan areas are expected to be in currently industrialized nations (Tokyo/Yokohama, New York/New Jersey, Los Angeles/Long Beach, and Osaka/Kobe). The smallest of these "top 20" is projected to exceed 10 million people. The new mega-cities of the developing world will require affordable infrastructure, and its development will place significant demands on already-strained national economies. New low-cost and cost-effective infrastructure technologies will be needed.

Evidence suggests that the scope of infrastructure will shift in the future as it has in the past. Telecommunications networks of several types, based on fiber optics and other "hard-wired" links or transmissions in various parts of the electromagnetic spectrum, are already part of the system but continue to expand; e.g., cellular telephone technology is becoming pervasive worldwide. Declining costs of tunneling, increasing land prices, and new technology (such as that demonstrated at M.I.T. by Tom Stockebrand) could foster growth of pneumatic tube transport systems for goods and passengers. Managed systems of parkland, open space, and wilderness may become elements of infrastructure, not only for their aesthetic and recreational value, but because their vegetation and ecologies could become essential parts of the supply cycles for clean water and air.

Empirical evidence[22] suggests that the time spent by individuals (on average) on transport is close to an anthropological constant: 1 to 1.5 hours per day, in all societies. This observation helps to explain why, as traffic congestion in central cities and high-activity suburban areas has grown severe, new urban centers have grown up on the periphery of older urban regions: people try to keep their travel time down to acceptable levels. If new technologies such as automated guidance systems for automobiles or more flexible and faster line-haul transit systems support higher-speed connections among cities and towns, we may expect further spreading of our urban populations across the countrysides.

Finally, new technology may change the way we manage our current infrastructure systems. For example, riboflavin (also known as vitamin B2) has been found to accelerate sunlight's ability to break down certain industrial pollutants in waste water, perhaps foreshadowing substantially improved waste treatment efficiencies. Advances in materials science are already yielding concrete with high strength - in tension as well as compression - and enhanced workability, that may enable us to build facilities that take less space and use less material. Genetically engineered algae and bacteria could allow sewage treatment to begin at the source - perhaps in tanks located next to the hot water reservoir in homes and commercial facilities, thereby reducing the load on central municipal plants.  Many other ideas could be suggested and will motivate researchers and infrastructure providers to find ways to enhance performance and profits.

## 4.5 TOMORROW'S INFRASTRUCTURE:  RESPONDING TO PROBLEMS AND TRENDS

Lessons from the development of old cities and the planning of new towns point the way toward the infrastructure of tomorrow.  While forecasting of technology and its use is risky business, the exercise is nevertheless a valuable way of encouraging exploration and perhaps shaping the future.  It is also irresistibly attractive to anyone who purports to be a "planner" and to have interest in the way cities are and might be.

The infrastructure of tomorrow will have to be more flexible than today's, able to respond more readily to shifting demands.  Under-used facilities represent a misallocation of investment and a drain on operating budgets, but facilities are in most instances taken out of service only when a competing mode can perform the service more effectively or because the service is no longer particularly valuable, for example, when a bridge is too narrow for increased traffic.  However, the abandoned facility remains in the landscape, sometimes distorting economic activity patterns and growing to be a hazard.

The conversion of infrastructure facilities to other uses - the term "adaptive reuse" is employed in real estate development - is one way to provide flexibility.  Abandoned railroad lines are being turned into biking and hiking trails in the United States, for example, and a few old water-powered mills have been converted to small--scale hydroelectric generation.  Today's highways could become tomorrow's high-speed lines.  Conversion through emplacement of new technology, i.e., "retrofit" will minimize disruption of existing community patterns.

The infrastructure of tomorrow will be environmentally more friendly than today's.  The public will demand it[23] and technology will enable us to deliver the goods.

A progressive internalization of environmental costs is an essential mechanism for making infrastructure environmentally more "friendly."  Efficient use of environmental resources would then become a crucial indicator of infrastructure performance.  Planning and design professionals are learning ways to achieve more effective integration of built and natural elements of the environment, and this will influence the scale and design of infrastructure. Recent plans for small-scale new communities in several areas throughout the United States demonstrate early steps in this direction.

We may come to accept standards of some aspects of service that are lower than in the past, for example, in terms of peak capacities in return for improved environmental or social impact. The widespread retention of a 55mph speed limit on major highways is an example of such acceptance. However, the scope of service will be enhanced overall.

The operation of tomorrow's infrastructure systems is likely to be more distributed than today's, i.e., there will be a better match between the locations where infrastructure's services are produced and where they are wanted. Improved treatment and recycling technologies will make this possible for waste treatment and water supply, and the trend is already apparent in telecommunications, with companies and individuals increasingly using their own equipment for directing and redirecting calls. For energy supply and transportation, new technologies need further development. For example, advanced photovoltaics, fuel cells, and batteries could permit power generation in home and office to supplement centrally generated electric power, or at least smooth the costly peaks and valleys of electricity demand. Electronic route guidance or palletized line-haul technologies may increase effective traffic flow capacity on major commuter routes, effectively converting private vehicles to higher capacity transit conveyances for part of their journey.

While the systems themselves may be more distributed, centralized and quickly responsive control will enhance overall performance. New electronic sensors and data management systems will facilitate continuous monitoring of facility operating conditions and adjustment of these operations. These control systems will give early warning of physical deterioration that requires maintenance, and can allow supply characteristics to be adjusted in response to demand. This latter capability is already available for much of the electric power grid (although loss of a link can still make the lights go out in large areas of a city) and is becoming more effective in vehicular traffic management, as more cities install coordinated signal systems.

"Privatization", the conversion to private provision of public services is a currently fashionable phrase for a management trend that could become increasingly a part of tomorrow's infrastructure. The more distributed nature of tomorrow's systems will foster this privatization by shifting part of the investment burden to individual households and businesses. In the public sector, procurement procedures with such acronyms as BOO and BOT (build-own-operate, build-own-transfer) will grow more popular as ways of mobilizing private initiative to supplement or replace government in provision of such infrastructure as toll roads and transit lines, that offer opportunities for adequate return on investment.

Sometimes the project is made viable by "packaging" it with other valuable rights, as was the case when land development in the 1920s played an important role in the growth of street railways in Los Angeles. Contracting with private concerns for maintenance, a popular privatization step in developing countries, is being studied closely by the Federal Highway Administration.

## 4.6 BUILDING THE INFRASTRUCTURE OF TOMORROW

Any such forecasts of future technology may seem - depending on one's point of view - impossible or inevitable. Which word is the best description depends greatly on the effectiveness of vision and leadership by those responsible for developing and adopting new technologies, for building tomorrow's infrastructure.

For example, Baron Haussmann, who became the prefect of Paris in 1853 and decisively influenced the shape of that city, saw to it that all Paris streets were equipped with sewers. The sewers came to house a range of other systems: two sets of water mains (one for drinking, the other raw Seine water for street cleaning and park irrigation), telegraph and telephone wires, pneumatic tubes for the postal service, tubes carrying compressed air, and later electrical systems for the traffic lights were suspended from the roofs of the galleries. Haussmann's vision helped turn the sewer system of Paris into "one of the engineering triumphs of the last century" and made it possible for infrastructure systems to evolve more easily.[24]

Today such vision is emerging from new understanding of our cities and their infrastructure. For example, we are coming to understand that metropolitan areas should be thought of (and managed) as functioning ecosystems. The potentially very productive role of undeveloped land - green space and urban wilderness - within that system should be used more effectively. In the 18th century, William Pitt, the first Earl of Chatham and a popular prime minister of England, asserted in a speech before the House of Commons that "the parks are the lungs of London," but we must move beyond that limited view. Environmentally friendly and compatible infrastructure - *ecostructure* - must be understood as the viscera, nerves, and circulatory organs of the city. The city of Arcata, California, for example, incorporates a wetland in its sewage treatment system.[25] Such uses of natural subsystems should be developed as part of our future water supplies, municipal waste management, and air pollution control.

This is a step (actually, several steps) beyond environmentally "friendly" infrastructure technology. The vision must be based on strong philosophical commitment to principles being defined under such rubrics as "sustainable" or "green" development. Just over two decades ago, the first Earth Day and passage of the U.S. National Environmental Policy Act were early accomplishments in developing these principles, and there is much yet to do. A transition to ecostructure is the goal.

**Fig. 4.3**     Community residents participated in construction of the Thomas Road Overpass on a Phoenix, Arizona interstate, literally making this potentially intrusive infrastructure an integral part of the community.

The 16th century English philosopher, Francis Bacon, wrote that "He who will not apply new remedies must expect new evils...." The infrastructure of tomorrow must deliver new remedies. The experience of old cities and new towns around the world point the way and, despite our best efforts, new evils will no doubt appear to supplement or replace today's urban problems. However, new infrastructure technology and - most important - people of vision to manage our technology will enable us avoid some problems, overcome others, and continue to deliver opportunities for a better life to all the world's people.

## Notes

1.  Straub, H. *Die Geschichte der Bauingenieurkunst.* Basle: Verlag Birkhauser, 1949. Tr. by E. Rockwell, *A History of Civil Engineering.* London: Leonard Hill Ltd., 1952.

2.  Herman, R. and Ausubel, J.H., "Cities and Infrastructure: Synthesis and Perspectives". In: J.H. Ausubel and R. Herman (eds.), *Cities and their Vital Systems, Infrastructure, Past, Present and Future.* Washington, D.C.: National Academy Press, 1988.

3.  Hamilton, S.B., "Building and Civil Engineering Construction." In: C. Singer, et al.(eds.), *A History of Technology, Vol. IV: The Industrial Revolution - 1750 to 1850.* Oxford: Clarendon Press, 1975.

4.  National Research Council, Committee on Infrastructure Innovation. "Infrastructure for the 21st Century." Washington, DC: National Academy Press, 1987.

5.  Reid, D. *Paris Sewers and Sewermen: Realities and Representations.* Cambridge: Harvard University Press, 1991.

6.  Lee, K S., Stein, J., and Lorentzen, J. *Urban Infrastructure and Productivity: Issues for Investment and Operations and Maintenance.* Washington, DC: World Bank, 1986.

7.  Aschauer, D. "The Macroeconomic Importance of Public Capital." Presented at a Colloquium: The Role of Infrastructure in America's Economy, sponsored by Financial Guaranty Insurance Company and The Public's Capital, Washington, D.C., 1989.

8.  Analyses conducted by the author.

9.  FCDA (Federal Capital Development Authority). *A New Federal Capital for Nigeria: Initial Draft Concept Plan.* Lagos, Nigeria, 1978; FCDA, A New Federal Capital for Nigeria: Preliminary Analysis of Cost and Financing. Lagos, Nigeria, 1979a; FCDA, *The Master Plan for Abuja, The New Federal Capital of Nigeria.* Lagos, Nigeria, 1979b.

10. Batam Centre Planning Group. *Batam Centre Master Plan*. Jakarta, Indonesia: Otorita Pengembangan Daerah Industri Pulau Batam, 1983.

11. Real Estate Research Corporation. *The Costs of Sprawl*. Washington, DC: Council for Environmental Quality and Environmental Protection Agency, 1974.

12. Pat Choate's *America in Ruins* in 1981 warned that U.S. public facilities wee wearing out faster than they were being replaced. There have been several subsequent studies: Congressional Budget Office, *Public Works Infrastructure: Policy Considerations for the 1980s*, 1982; National Infrastructure Advisory Committee, *Hard Choices*, 1984; National Council on Public Works Improvement, *Fragile Foundations*, 1986; Congressional Budget Office, *New Directions for the Nation's Public Works*, 1988; Office of Technology Assessment, *Delivering the Goods: Public Works Management, Technology, and Financing*, 1991.

13. World Bank. *Sub-Saharan Africa: From Crisis to Sustainable Growth*. Washington, D.C., 1989.

14. Grubler, A. *The Rise and Fall of Infrastructures: Dynamics of Evolution and Technological Change in Transport*. Heidelberg: Physica-Verlag, 1990.

15. Marland, G. and Weinberg, A., "Longevity of Infrastructure." In: J. Ausubel and R. Herman (eds.), *Cities and their Vital Systems*. Washington, D.C.: National Academy Press, 1988.

16. White, D.C., Andrews, C.J., and Stauffer, N.W., "The New Team: Electricity Sources Without Carbon Dioxide," *Technology Review*, January 1992, pp. 42-50.

17. Hanks, J.W., Jr., and Lomax, T.J. *Roadway Congestion in Major Urban Areas 1982 to 1988*. College Station, Texas: Texas Transportation Institute, 1990.

18. Cannon, J.S. *The Health Costs of Air Pollution: A Survey of Studies Published 1984-1989*. New York: American Lung Association, 1989.

19. Author's estimate, based on distributions of U.S vehicle and highway mileage.

20.  Hanson, op. cit. 1988.

21.  Habitat (United Nations Centre for Human Settlements).  "Global Report on Human Settlements."  Oxford: Oxford University Press, 1987.

22.  Zahavi, Y. *Travel Characteristics in Cities of Developing and Developed Countries.*  Washington: International Bank for Reconstruction and Development, 1976.

23.  For example, a national survey conducted by Yankelovich Clancy Shulman, as reported in The Wall Street Journal in September 1991, showed that public attitudes strongly favor stiffer fines for corporate water polluters, perhaps with prison terms for offending company executives.

24.  Reid, D.  *Paris Sewers and Sewermen: Realities and Representations.* Cambridge: Harvard University Press, 1991.

25.  Spirn, A.W.  *The Granite Garden: Urban Nature and Human Design.*  New York: Basic Books, 1984.

# 5

# Inland Transport in Europe - Trends and Prospects[1]

**Joseph Elkouby**
*Ingénieur-Général des Ponts-et-Chaussées, Paris*

## 5.1 INTRODUCTION

In February 1986, a European Agreement (called *Acte Unique*) was signed between the twelve countries of the European Economic Community with a view to strengthening the cooperative efforts among the countries. The first of January 1993 was set as the ultimate date for a truly integrated market in which national boundaries and other remaining obstacles to trade and industry should have been eliminated. As the final objective of this European Agreement was to foster economic efficiency and facilitate the social integration of all citizens of the European Community, several studies were undertaken to shed light on the major aspects of the unification process. The improvement of transport systems between and within the member countries appeared as an essential element of this process, not only because of the advantages expected for trade and travel, but also because in the long run it would influence the development of new activities.

I would like to present an overview of the transport studies carried out within the European Community. I shall try to describe the main components of transport demand that can be foreseen in the next twenty years (that is, at the horizon 2010) and the infrastructures needed to cope with them. My explanations will be mainly centered on France (as it is the case I know best), and also because it is truly at the crossroads of the major traffic flows between the European member countries.

This article centers on three aspects:

- the changing socio-economic system, and in particular the requirements resulting from general as well as European objectives;
- the major trends of the evolving transport demand and travel needs between regions, cities, or economic entities; and
- the infrastructure networks needed with an indication of their priorities.

## 5.2.    THE IMPLICATIONS OF A EUROPEAN APPROACH

### 5.2.1    The Geographic Constraints

One of the primary goals among the "founding fathers" of the European Economic Community (EEC) was to create between the United States of America and the Eastern Bloc a large area of economic development and prosperity through the elimination of national boundaries and common use of basic resources:  coal, steel, agriculture, nuclear power, and transport.

However, in the transport sector, the European dimension was not for many years taken into account.  In fact, in 1985 the European Court of Justice criticized the Council of Ministers for the lack of a coherent overall transport policy which should have been formulated to better coordinate transport investments and operations in the member countries.

The current transport system actually consists of twelve national systems which are, fortunately, linked together; differences remain, however, between countries as to legislation and control procedures while several links lack sufficient capacity to accommodate international traffic, or simply do not exist.

When considering the territory of the EEC, a first remark is in order: compared to any individual member country, the EEC introduces a change in scale in terms of area, population, and more specifically, distances.  For instance, the EEC covers four times the area of France alone and has a total population six times that of France.  Transport distances which could hardly reach 1,000 km. when confined by national boundaries, would now be in the range of 2,000 to 3,000 km, perhaps even more if new countries join the Community.

The elimination of national boundaries also emphasizes the role played by the area sometimes called "the European Spine", i.e., an area of high population density extending from London to northern Italy through the north and east of France (see Figure 5.1); this area includes several large urban agglomerations:

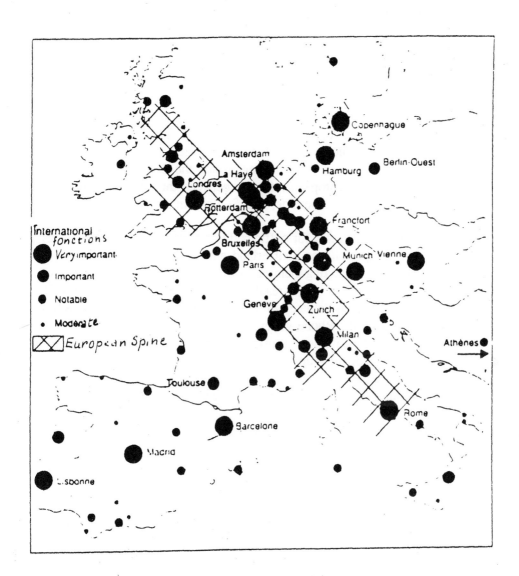

**Fig. 5.1 Map of Major Urban Agglomerations in Europe (*Le Monde*, 4 March, 1988).**

London, Amsterdam, Rotterdam, Brussels, Lille-Roubaix-Tourcoing, Strasbourg, Mulhouse, Bale, Dusseldorf, Cologne, Frankfurt, Mannheim, Stuttgart, Zurich, and Torino (70 million inhabitants, compared to the megalopolis of the East Coast of the U.S. with approximately 45 million inhabitants).

For the largest portion of its territory, France is located outside of this densely populated area, even though some border towns are included in it. Thus, the Paris metropolitan area which stands aside, acts as a counterweight protecting France from the disadvantage of a marginal position. A first conclusion should be drawn from this geographic analysis: the obvious need for France to provide efficient transport links between the Paris region and the European "megapole" and strengthen its role as a European turntable between the UK, Germany, and the Scandinavian countries on one side, and South European countries on the other.

In this function as passageway for traffic between north and south, France has a major role to play, especially as Switzerland and Austria (which are not yet members of the EEC) are protecting themselves from through-traffic by restrictive measures on truck movements. As a result, a large proportion of truck traffic between Germany and Italy is deviated to the west and may increase congestion of the highway system in the Rhone valley.

At the same time, France's decentralization policy, which started some ten years ago, has given a political status to the 22 regions. Therefore, the response to a changing economic base and new transport needs cannot be expected, as was the case before, from a powerful and centralized administration. The solutions and, in particular, decisions on the main transport infrastructures and their financing, require a preliminary agreement between the national and regional authorities involved. With the elimination of national boundaries, some regions of France have developed a permanent relationship with neighboring regions beyond French borders and are cooperating and supporting together projects of common interest. However, despite many examples of such cooperation, there remains in Europe, on the whole, a fairly strong competition between large cities and regions which are trying to attract major industrial corporations and their high-level jobs.

## 5.2.2   The Changing Socio-Economic Background

The close relationship between economic development (which is the basis of social progress) and the performance of the transport system has often been emphasized. In all Western countries, the transport system plays a central role, because transport operations are included in many industrial processes and are a key factor for commercial enterprises.

In Europe, transport demand has substantially increased over the past twenty years and will continue to grow in order to respond to the desires of the population in terms of speed, safety, and comfort. However, for industrial and commercial activities, reliability, more than speed, will be the key element shaping the demand for transport services.

While transport users benefit from a better performance of transport systems and from infrastructure improvements, the average citizen is more and more concerned about the negative impact of transportation on his natural environment as well as on his daily life.   Other activities such as industry, housing construction, and agriculture also have environmental impacts.  But the transport sector, especially because of the increasing availability of the private automobile, has led to an explosion in mileage travelled in Europe, and has resulted in a dramatic impact on the ecological system of many regions.  If the resulting pollution is not curbed, this may raise global issues and seriously affect living conditions of future generations.

The major threat is due to air pollution resulting mainly from vehicle emission gases.   Despite a number of tentative control measures, and the increasing use of unleaded fuels, the problem of air quality is far from an acceptable solution.  Of even greater concern is the famous "greenhouse effect" and the related global warming with its possible consequences on climate change. All vehicles powered by internal combustion engines, that is, road vehicles, use fossil fuels and are thus responsible for $CO_2$ emission, the main greenhouse gas.

All European countries report that $CO_2$ emission levels from road traffic have substantially increased over the past ten years.  There is hope to reduce these emissions by technical measures (for example, improved fuel efficiency, hybrid electric cars, driver's education).  But the progress will be slow and the reduction gained in the amount of gas pollutants may well be offset by the expected increase in traffic.

It is clear that because of its environmental impact and other adverse effects, road traffic in Western Europe cannot continue to grow at the same pace as experienced over the past ten or twenty years. The principles suggested for "sustainable development" will impose a limit on the growth of motor traffic, and especially on the increasing number of private automobiles. In the United Kingdom, for instance, from 1963 to 1987, i.e., in 24 years, travel has more than doubled, totalling some 600 billion person/km, out of which 84% is by car. Based on a study carried out in 1989, the traffic forecasts predict a further increase of total road traffic between 83% and 142% by the year 2025, if no specific measure are taken to control the demand, and assuming, of course, that motorway and road infrastructures have been extended in due proportion to meet traffic demand. Similar trends can be foreseen in other member countries such as France, Belgium, and Germany where the rate of car ownership per household is higher than in the UK.

One can see that the trends in travel and traffic growth are in conflict with the aims and means of achieving sustainable development. As already pointed out, road traffic is a major user of fossil fuels which are non-renewable energy resources. The shift to renewable energy, or to electric power in cars, will only have a marginal effect on this travel mode. The best means toward a solution will combine the reduction of travel needs (through various regulatory measures) and a shift to public transport, particularly rail transport systems which rely almost exclusively on electric power in Western Europe.

However, despite this obvious advantage of rail transport over road traffic, some ecology-minded groups still oppose the construction of additional TGV links, as they often successfully defeated new motorways in the same corridors. This reluctance usually comes from population groups living near the new infrastructures who fear the deterioration and damage inflicted upon their landscape and way of life; they show little understanding of the so-called "macro-economic" benefits expected for their region or the country as a whole.

These groups feel they are sacrificed to the modernization of transport or to the well-being of transport users, and their opposition to the construction of a single link may jeopardize the coherence and operation of the whole transport network. Such conflicts, it is hoped, will be solved through improved decision-making procedures based on better and more impartial distribution of information to citizens' groups and also on a more efficient dialogue between state and regional authorities.

Besides the pollution problem, another aspect of the transport function, mainly in industrialized countries, is its role in the process of industrial production. This is usually referred to as the "logistic revolution", inasmuch as the new industrial processes tend to optimize transport operations through comprehensive information and tele-transmission techniques. More and more, transport has become an integrated part of production and distribution systems used by industrial and commercial firms.

This also explains the leading role of the trucking industry which makes it possible to organize door-to-door services with minimal delays. Consequently, industrial firms tend to operate with stocks kept at a minimum level (just-in-time supply system). This principle is also valid for major distribution firms (such as shopping centers or department stores) which rely on a stable supply system associating efficient transport operations with reduced warehouse space. In such a system, the controlling factor is no longer the price or duration of transport, but the dependability of the supply system as a whole.

Against this background, one can easily understand why the location of a new plant should be determined only after a careful screening of the highway links needed for its supplies. Other transport factors such as airline and railway services should also be taken into account, to respond appropriately to the travel needs of their professional and executive groups. More generally speaking, today's corporations, which are usually concentrating on high-tech products, are rather mobile, and thus especially looking for locations with the best combination of transport links and amenities.

This approach to industrial locations finally results in a fierce competition between various cities or regions, not only within France, but in the whole European territory. There is obviously a risk that existing population centers and best-equipped regions in Europe continue to attract the new firms with their employment potential and lead to the depletion of the less-developed areas. This is, no doubt, a challenge that will require the definition of a regional development policy on a European scale with a view to promoting more balanced development between regions and, at the same time, improving their transport infrastructure as well as their environment.

## 5.3 TRENDS IN TRANSPORT DEMAND

The primary objective of transport infrastructure is to provide for the safe and efficient movement of people and goods. Thus, a planning approach is necessary in order to forecast future transport needs, starting from assumptions of population growth and economic activity and also taking into account the level of service requested by transport users and environmental factors. Throughout history, experience has shown that transport activities, i.e., transport demand and supply, have been an essential part of economic growth which, in turn, calls for an increase in transport. On the other hand, progress in transport efficiency provides access to a wide range of economic resources, thus spurring economic activity.

Considering the EEC, it should be noted that for the past twenty years, the member countries have more or less followed similar policies in transport matters in order to facilitate communications between each other and give the users a certain freedom of choice as to transport modes. As a result of bilateral agreements and technical coordination measures, the development of passenger and goods traffic has roughly followed the rate of growth of GDPs (see Figures 5.2 and 5.3).

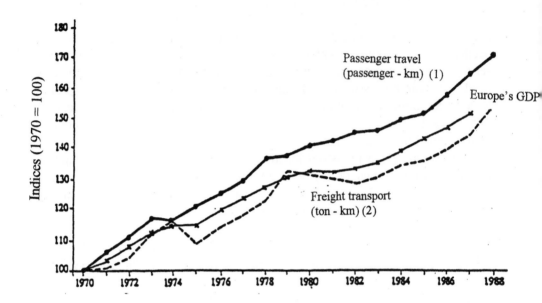

**Fig. 5.2  Trends in Passenger Travel and Freight Transport in Europe, 1970-1988.**

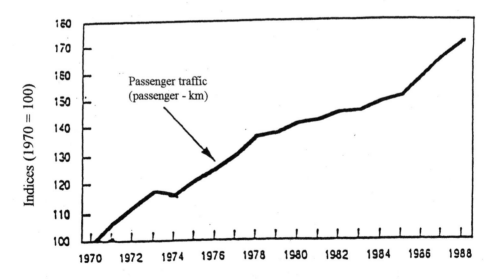

**Fig. 5.3  Trends in Passenger Traffic, 1970-1988.**

For example, Figure 5.2 shows that for Europe as a whole (17 countries), total passenger travel and goods transport have developed at a pace comparable to that of GDP. One can notice a slight recent increase in passenger travel and some changes of trend in the curve of goods movement in 1974 and 1979 which correspond roughly to the years of the two oil crises. The figure shows that the impact of the oil crises was stronger on goods than on passenger traffic.

In order to assess the level of future transport demand for the year 2010, it was necessary to select a growth rate of GDP; ultimately, a rate of 3% per year was retained for France and neighboring countries. This assumption may sound rather optimistic, but appears consistent with the experience of the past twenty years. On that basis, an overall growth rate for the transport sector may be derived; however, the forecasts should be adjusted in each category (passenger or freight) with reference to past traffic trends and requirements for an improved quality of service.

### 5.3.1 Evolution of Passenger Flows

Under the assumption of a reasonable rate of economic growth in Europe, one can expect an increase in the total number of passenger trips as well as an increase in the average trip length. It has been assumed that between 1990 and 2010, trip length increase would be equivalent to that observed from 1970 to 1990. This assumption can be justified on several grounds: rise of living standards, increased leisure time, and growing number of passenger cars. At the same time, the development of high-speed transport and the elimination of national boundaries will stimulate exchange flows between countries, thus encouraging longer trips.

The assumptions above have led to an overall forecast of transport demand, all modes included, at the horizon of 2010 comprised between 1.8 and 2.2 times the 1990 traffic, that is, roughly a doubling of current transport flows. This, of course, should be understood as an average rate which does not apply rigidly to each mode. Adjustments should be made in each corridor according to trip length, geographic links, purpose of trip (work, visit, leisure, vacation, etc.). In fact, transport planners will be confronted with a diversification of transport demand, due to several factors, such as:

- a larger share of senior citizens among transport users, greater mobility of young families, and development of short and frequent leisure trips; and

- heavier impact of the time factor, as more and more users wish to reduce journey time; hence, the preference of travelers for the use of airlines, motorways, and high-speed rail and the need for transport planners to minimize the overall journey time, improve quality of service and dependability of the transport system including all modes.

### 5.3.2 Modal Distribution of Passenger Traffic

Besides the above factors which describe the social and economic characteristics of transport users, the share allocated to each mode is highly dependent on the effective transport capacity available. For instance, the introduction of high-speed rail services has substantially changed the distribution of passenger flows between airlines and rail, but has also affected, perhaps less dramatically, the respective share of rail and motorway traffic (see Figure 5.4).

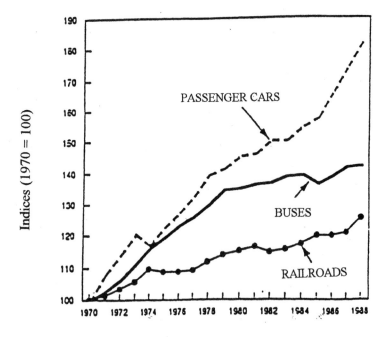

Fig. 5.4  Passenger Travel in Europe Modal Distribution, 1970-1988

It can be noted that:

- In Europe the use of passenger cars is now so widespread that passenger car traffic (expressed in person/km) is largely predominant (from 71% to nearly 80% in 1990); bus travel has maintained its share over the past twenty years, while rail traffic has decreased by roughly one-third.
- In France, a similar evolution occurred; however, the share of rail traffic is slightly higher than in Europe as a whole (10% versus 8%), probably because of the high-speed stretch between Paris and Lyon.

Therefore, to assess the twenty-year traffic forecasts, the demand was broken down into three categories:

1) inter-urban trips of medium length (less than 200 km);
2) long distance inter-urban trips (more than 200 km);
3) trips within urban or metropolitan areas.

For trips of the first category, we do not foresee any technical innovation capable of substantially reducing the predominant use of the passenger car. Because of its obvious advantages (availability, convenience, and flexibility), the passenger car will maintain its share of total traffic and might even expand it.

For long-distance trips, the modal distribution will certainly be altered by the continuing development of high-speed rail transport and also by a better organization of the "transport chain" (through improved access to rail stations and airports and easier connections).

Finally, concerning urban and suburban traffic (representing about 30% of passenger trips), the proper solution will be found in a better balance between the use of the private car and mass transit.


### 5.3.3   Evolution of Freight Transport

Both passenger traffic and freight transport depend very much upon overall performance of the economy and on the cost of energy.   However, there is a greater correlation between freight movement and the behavior of major economic entities such as large industrial and commercial firms as well as major carriers. As a result of efforts launched by these economic "actors" to reorganize manufacturing and distribution activities at the European scale, and also because of the increased utilization of containers in world transport, the inland transport system has gradually changed to adapt itself to the new transport needs.

Four major trends are at work shaping this new demand:

- an increased trade of manufactured and intermediate products and other varied goods (usually high-value products) and the reduction of bulk transport;

- in the manufacturing sector, greater concentration of and more specialized plants, relying more and more on just-in-time processes;
- the distribution of goods (for manufacturing or commercial purposes) operating with minimal stocks which are concentrated in selected locations serving one or more large regions; and
- manufacturers tend to entrust some specialized firms with "logistic tasks" so such firms could introduce economies of scale in transport and lead to a more rational management of manufacturing inputs. In some cases, this evolution may end up with the construction and operation of "logistic centers" designed to serve a whole region of Europe, or even several regions.

During the past twenty years, the growth of freight transport has been quite different among the members countries (see Figure 5.5). In northern Europe, where trade was already active, the increase was around 30% (slightly less for France). In southern European countries, a higher increase was observed (almost double the 1970 level).

The forecasts over the next twenty years are difficult to assess. In the case of France, for example, such forecasts are directly dependent upon the growth of its manufacturing industries and the sale of its products on the European and world markets. However, they may also be influenced by other factors such as the opening of the Channel Tunnel (thus generating increased through-traffic between north and south Europe), the success of development policies in countries of southern Europe, and the reorganization of freight handling in French ports.

For Western Europe as a whole, on the basis of projections put forward by several experts, it is expected that the overall freight movements within and between member countries will be 1.4 to 1.6 greater than the 1990 figures. This should be considered as an average growth factor. The increase may be much higher in specific cases, in particular for:

- routes connected with major ports (Marseille, Le Havre, and Seine Valley);
- routes linking the Channel Tunnel to Italy via the Rhone Valley, or to Spain via the southwest region;
- routes from Germany to Spain (via the Rhone Valley and Languedoc-Roussillon).

(billions ton/km)

| Country | Traffic, 1970 | Traffic, 1988 | Percent increase 1970-1988 |
|---|---|---|---|
| Germany | 197.31 | 262.75 | 33% |
| France | 148.07 | 171.43 | 16% |
| United Kingdom | 111.50 | 145.31 | 30% |
| Netherlands | 46.84 | 59.18 | 26% |
| Belgium | 27.70 | 37.54 | 36% |
| Denmark | 9.65 | 10.75 | 11% |
| TOTAL | 541.07 | 686.96 | 27% |
| Italy | 77.12 | 183.73 | 138% |
| Spain | 62.04 | 145.00 | 134% |
| Greece | 7.65 | 13.66 | 79% |
| Portugal | 5.00 | 10.05 | 101% |
| TOTAL | 151.81 | 362.44 | 132% |
| Ireland | 3.50 | 5.52 | 58% |
| Luxembourg | 1.20 | 1.27 | 6% |
| TOTAL EEC | 697.58 | 1,046.19 | 50% |
| Switzerland | 10.92 | 15.95 | 46% |
| Austria | 14.03 | 19.47 | 39% |
| TOTAL | 722.53 | 1,081.61 | 50% |

**Fig. 5.5   Trends in Freight Transport in Europe, 1970-1988.**

## 5.3.4   Modal Distribution of Freight

The just-in-time procedures mentioned earlier, which are applied to manufacturing and distribution activities do not necessarily mean fast transport services. For the majority of goods concerned the just-in-time concept refers basically to security of supply.   More than speed of transport proper, what is important is the dependability of the whole transport system together with its flexibility.   In addition, electronic transmission of computerized data will be more and more utilized, thus improving logistic efficiency.

Outside of small parcel deliveries and high-value goods, inland freight is accommodated in various proportions by road, rail, waterways, and pipelines. During the past twenty years, as observed for passenger traffic, road services are predominant, with a strong tendency to increase their share of the transport market (see Figures 5.6 and 5.7 and 5.8). It may be noted that:

- the market share of road haulage in the total of freight in northern Europe has jumped from 49% in 1970 to 65% in 1988, an increase of 33%;
- rail has lost 37% of the freight market over the same period of time (but only 33% in France);
- the waterway share has been less reduced than the freight on rail (22% in northern Europe); and
- the use of pipelines has substantially increased, substantially replacing rail and water transport.

| Country | ROADS | | | RAILROADS | | | WATERWAYS | | |
|---|---|---|---|---|---|---|---|---|---|
| | Share 1970 | Share 1988 | % incr. | Share 1970 | Share 1988 | % incr. | Share 1970 | Share 1988 | % incr. |
| Germany | 40% | 58% | 46% | 36% | 22% | 38% | 25% | 20% | -19% |
| France | 45% | 65% | 46% | 46% | 11% | -13% | 10% | 4% | -55% |
| U.K. | 76% | 86% | 13% | 22% | 13% | -43% | 2% | 2% | -12% |
| Netherlands | 25% | 37% | 41% | 8% | 5% | -12% | 66% | 57% | -13% |
| Belgium | 47% | 67% | 42% | 28% | 19% | -32% | 24% | 14% | -44% |
| Denmark | 81% | 84% | 5% | 19% | 15% | -19% | 0% | 0% | -10% |
| TOTAL | 49% | 65% | 33% | 33% | 21% | -37% | 19% | 15% | -22% |
| Italy | 76% | 89% | 17% | 23% | 11% | -55% | 0% | 0% | -83% |
| Spain | 83% | 92% | 10% | 17% | 8% | -50% | 0% | 0% | -57% |
| Greece | 91% | 96% | 5% | 9% | 4% | -51$ | 0% | 0% | -44% |
| Portugal | | 85% | | | 15% | | | o% | |
| TOTAL | 80% | 90% | 13% | 20% | 10% | -52% | 0% | 0% | -82% |
| Ireland | | 90% | | | 10% | | | 0% | |
| Luxembourg | 12% | 22% | 89% | 63% | 50% | -20% | 25% | 28% | 13% |
| TOTAL EEC | 55% | 73% | 33% | 30% | 17% | -44% | 15% | 10% | -35% |
| Switzerland | 38% | 52% | 37% | 60% | 47% | -22% | 2% | 1% | -40% |
| Austria | 20% | 33% | 64% | 70% | 58% | -18% | 9% | 9% | -2% |
| TOTAL | 54% | 72% | 34% | 31% | 18% | -42% | 15% | 10% | -34% |

Fig. 5.6 Freight Transport in Europe, Modal Distribution, 1970-1988.

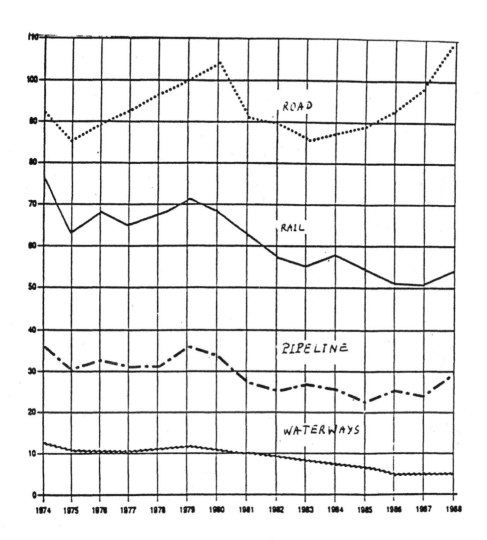

Source : OEST

Fig. 5.7  Modal Distribution of Freight Transport in France, 1974-1988

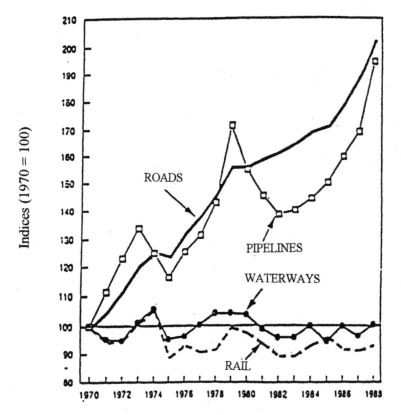

**Fig. 5.8 Growth in Roads, Pipelines and Waterways**

**source** ECMT

### 5.3.5    Major Transport Flows in Europe

In the European Community transport between member countries represents a substantial part of the total movement and is bound to increase again in the future with the elimination of border controls. This trend should be taken into account in the design of infrastructure networks, as a better quality of transport services on international links is a good way to accelerate European integration.

Concerning passenger flows, there are no statistics available on international travel by road (except for buses). On the rail network, international passenger traffic is about 12% of the total rail traffic of the EEC and shows a rather slow rate of increase, less than domestic rail traffic. This demonstrates that there is still some border effect which will fade off gradually in the future.

As to freight, international flows between member countries account for about 20% of the total of inland freight observed in the 12 countries of the EEC. France's share alone (25%) is higher than the average proportion observed, because of the weight of through-traffic between north and south Europe. These exchange flows between member countries were growing at the same rate as domestic flows until 1986; since then the rate of increase has markedly changed, probably as a result of the integration of Spain and Portugal into the EEC. Between 1981 and 1987 the exchange flows between Italy and the rest of the EEC have increased by 12% and those with Spain by 74%. Moreover, international flows, mostly long distance traffic, usually concentrate in the best-equipped corridors. For example, 40% of the freight passing through the Rhone Valley consists of international trade and half of this proportion has its origin or destination in Spain or Italy.

## 5.4 INFRASTRUCTURE NETWORKS

### 5.4.1   The Highway Network

Until World War II, France was very proud of its national highway network, totalling 80,00 km; it felt, however, the need to build motorways on specific links in the years 1955-1960 and decided only in 1965 to launch a large construction program of inter-urban motorways with funding based on toll revenues. This program became necessary because of the continuing development of motorization and the steep increase of road traffic.

While this first motorway program was implemented (to ease congestion on major routes such as Lille-Paris-Lyon-Marseille, Bordeaux-Toulouse, etc.), another road improvement scheme was prepared for other important routes on which tolls were not economically justified (for instance, improved road network in Britanny, National Road No. 7). This policy, which advocated gradual improvement of road infrastructure to avoid the large capital investments involved in toll motorways, was criticized because of several drawbacks: lower safety standards, average speed lower than on motorways, and expensive improvement costs.

The experience gathered from the above programs led national authorities to the definition of a four-tier network:

1.   the motorway network, which includes all major inter-urban links, with dual carriageways and complete control of access; this network is designed to accommodate medium- and long-distance traffic in optimum conditions of speed and safety;

2.   the intermediate network, which includes all inter-urban routes on which full motorway features are not deemed necessary (network for regional or inter-regional traffic);

3.   a large network of service roads;

4.  the urban road network, designed to accommodate large flows of traffic
    at reduced speed.

This policy of specialized tiers is likely to be adopted by other member
countries with the objective of obtaining comparable design characteristics in the
EEC; this is especially important for the first two categories, on which we will
focus our attention.

We have already seen that total traffic is bound to double in twenty years and
that roads will take a major part of it. However, when considering the territory
of France, it should be stressed that:

- on the 7,300 km.of motorways in existence in 1990, half already bear a
  high traffic load (traffic congestion experienced during a large number of
  days annually, particularly at weekends or vacation periods);
- on the main north/south route, the percentage of heavy trucks is currently
  around 30%; and
- on the national road network (34,000 km, including motorways), 21%
  suffer from frequent congestion.

Another drawback of the existing motorway network is the fact that it is designed
around Paris as the center of a "spider's web"; outside of the Calais-Reims link,
there is no transverse motorway route. From this viewpoint France still lags
behind northern European countries such as The Netherlands and Germany where
the network is conceived more like a grid.

Therefore, a long-term road improvement scheme, known as the Master Plan
for the National Road Network, has been approved by the French government.
This scheme has the major goal of a new road investment program which is to
extend the total length of the motorway network to 12,1200 km., i.e., 9,500 km.
of actual motorways and 2,600 km. of roads with intermediate characteristics.


## 5.4.2   The Railway Network

As mentioned earlier, the freight carried by rail has been in sharp decline over the
past twenty years, not only in terms of market share of total ground
transportation, but in some countries in absolute value. And the main beneficiary
of this loss is the trucking industry. France remains, however, the country in
Europe where the railway system offered a significant resistance to this traffic
evasion.

In the course of transport studies conducted in France, a detailed analysis has
been launched to identify the causes of such a decline and find solutions for at
least stabilizing the volume of freight handled by rail. This is an important
objective on a few specific transport corridors in which truck traffic on

motorways reaches frequent congestion levels and creates hazards for non-commercial vehicles.

Besides a number of measures aimed at better management of freight trains on the rail network and a more expeditious handling of goods at transfer stations, some consideration has been given recently to the concept of combined rail and road transport, either with the use of containers carried on rail and transferred onto trucks for final delivery, or with piggy-back type systems. Studies are now underway to assess the portion of traffic that could utilize combined transport on major corridors such as the Rhone Valley. The EEC is also showing an increasing interest in the potential offered by combined transport. It would take too much time to mention the other measures and incentives that would be needed to reach a better balance between rail and road in moving and handling freight.

### 5.4.3    High-Speed Rail Services

Now we will turn to the more innovative role of railways in the field of passenger traffic as a result of the introduction of high-speed rail links. It has been pointed out that passenger travel rail services have experienced a reduction of their market share over the past twenty years while at the same time the highway system and airlines were gaining in patronage. This trend has been noted in all member countries of the EEC. However, this loss in market share has been smaller in France thanks to the development of high-speed rail links and the improvement of suburban and regional rail services. The experience gathered from the South-East TGV, the first high-speed link 417 km. long between Paris and Lyon, has demonstrated that on this range of distances, TGV can easily compete with other modes such as highways or airlines.

Since the construction of the Japanese Shinkansen and then the French TGV, most developed countries have become interested in high-speed rail transport because this innovative system substantially reduced journey times and used electric power, thus causing less damage to the environment.

Finally, the solution adopted in France is based on the construction of a high-speed rail network especially designed for high-speed trains; at the same time, existing rail infrastructures have been improved in order to accommodate high-speed trains at intermediate speed (i.e., 130 to 180 km/h), enabling passengers to reach a large number of destinations without transfer. A good example is given by the South-East TGV serving, besides Lyon, cities such as Geneva, Chambéry, Annecy, Saint-Etienne, and even Marseille. In other terms, the great advantage of the French high-speed trains is the fact that they can run on specific high-speed tracks as well as on improved existing railways. In fact, both networks remain compatible, thus making the large investment of new high-speed infrastructure economically viable (see Figures 5.9 and 5.10).

At the request of the French Minister of Transport, the French Railway Corporation (SNCF) initiated various studies with the objective of establishing a kind of master plan of high-speed rail services. However, because of the economic range of TGVs (500-800 km.), this master plan requires that it be coordinated with similar projects of neighboring countries (for example, north TGV and connection to the Channel Tunnel and British Rail). Hence, the idea of a European TGV scheme. This scheme was prepared in 1990 at the request of the European Commission and in association with the Transport Ministers of the 12 EEC member countries, for the year 2010.

At the same time, this first step of the EEC drew the attention of other European countries outside of the EEC and led to the establishment of an extended European scheme, including central and eastern Europe and the Scandinavian countries (see Figure 5.14). The scheme was prepared by a working group organized within the International Union of Railways and was presented in April 1992 in Brussels at the Eurorail Conference. It included three types of rail infrastructures:

- completely new lines especially designed for high- or very high-speed (300 km/h and even 350 km/h);
- improved existing lines permitting speeds up to 200 km/h;
- other existing lines used for distributing or collecting traffic from various origins.

In the perspective of the year 2010, projects under consideration in the EEC as well as Switzerland and Austria may total about 12,000 km. of new high-speed lines. If we limit our horizon to the year 2000 and on the basis of decisions already taken, we can foresee a network of 3,100 km. of new lines in Western Europe. This means that 1,200 km. will be added between 1993 and 2000 to the 1,900 km. in operation through 1992 (see Figures 5.11 and 5.12).

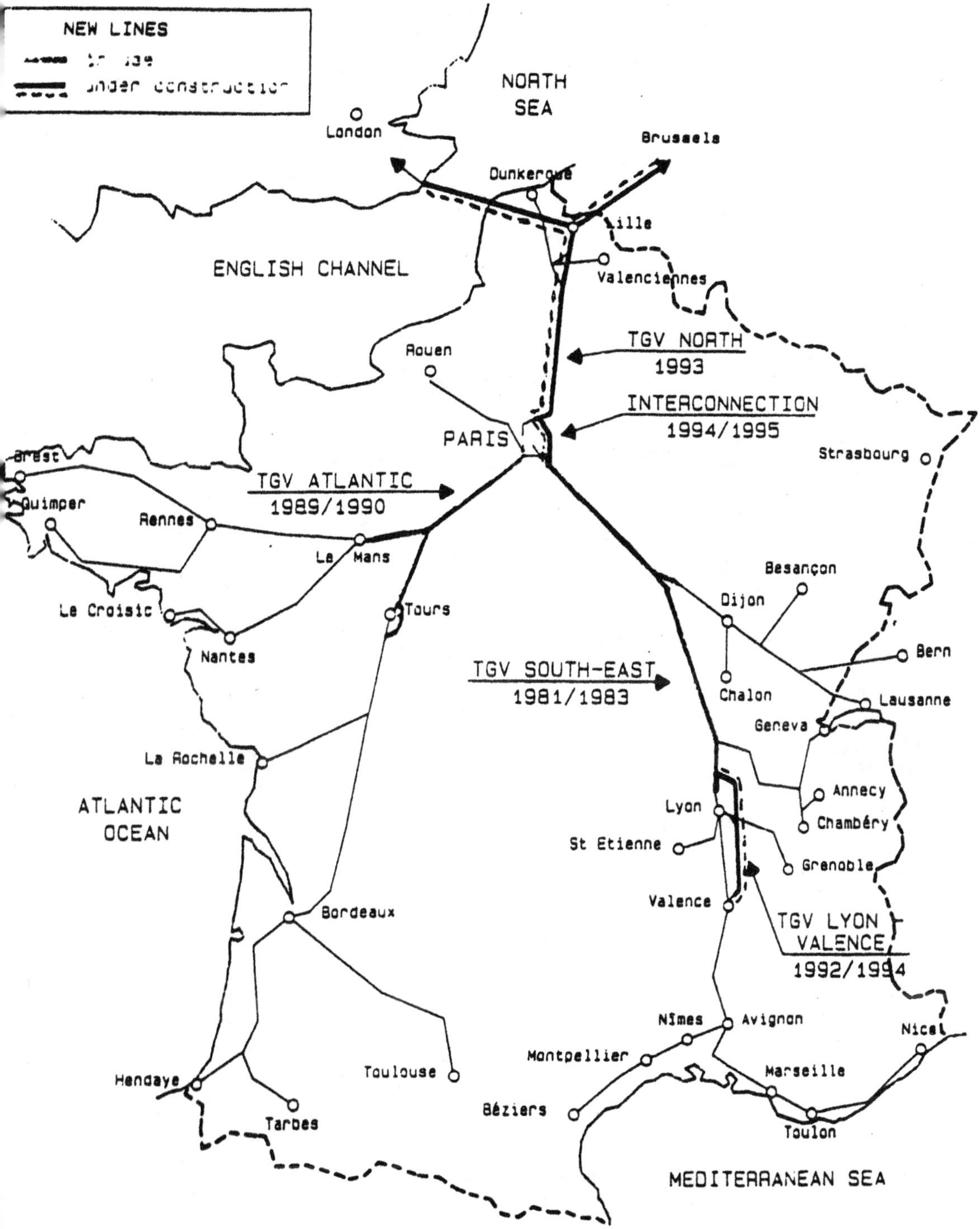

**NEW LINES**

in use

under construction

NORTH
SEA

London

Brussels

Dunkerque

Lille

ENGLISH CHANNEL

Valenciennes

Rouen

TGV NORTH
1993

INTERCONNECTION
1994/1995

PARIS

Strasbourg

TGV ATLANTIC
1989/1990

Brest

Quimper        Rennes

La Mans

Besançon

Dijon

Bern

Le Croisic        Tours

TGV SOUTH-EAST
1981/1983

Chalon

Lausanne

Nantes

Geneva

La Rochelle

ATLANTIC
OCEAN

Lyon

Annecy

St Etienne

Chambéry

Grenoble

Valence

TGV LYON
VALENCE
1992/1994

Bordeaux

Nîmes        Avignon

Nice

Montpellier

Marseille

Hendaye

Toulouse

Béziers

Tarbes

Toulon

MEDITERRANEAN SEA

**Fig. 5-9.  TGV Network in Use in 1995.**

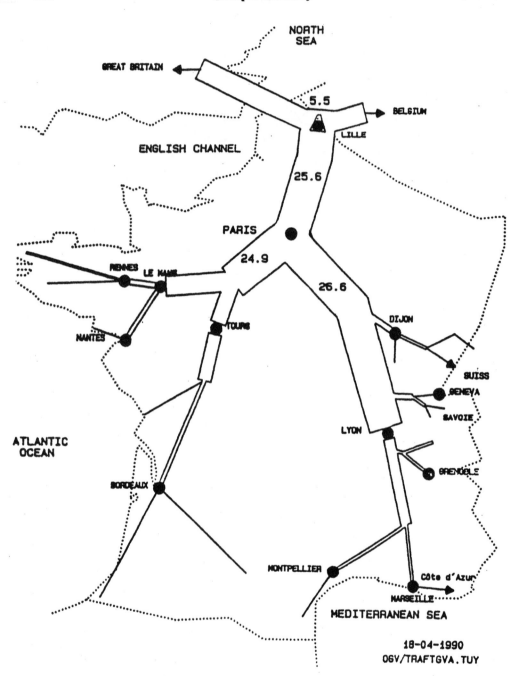

**Fig. 5-10.  French High-Speed Network Traffic for 1995.**

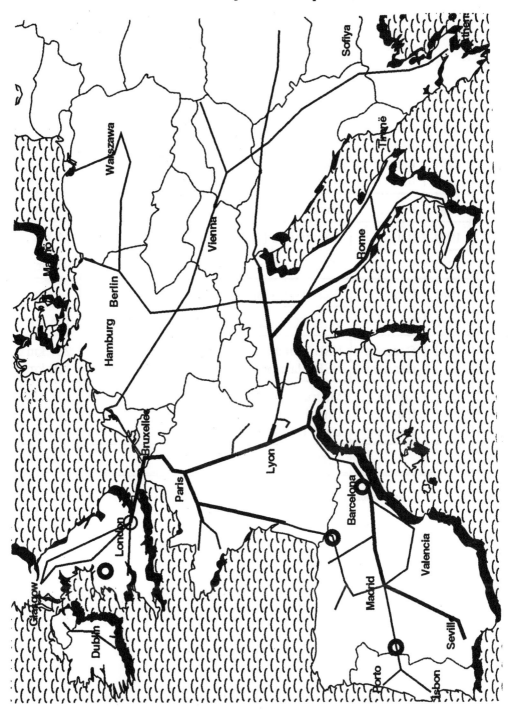

**Fig. 5.11  TransEuropean High-Speed Network** (reproduced by permission, from *Le Tunnel sous la Manche* by Bertrand Lemoine, *Le Moniteur*, Paris, 1994)

(new lines in operation, end of 1992)

| 1981 | France | South-East TGV (1st section) | 301 km |
| | Italy | Direttissima (1st section) | 150 km |
| 1983 | France | South-East TGV (2nd section) | 116 km |
| 1984 | Italy | Direttissima (2nd section) | 74 km |
| 1988 | Germany | NSB Wurzburg-Hannover (1st section | 90 km |
| 1989 | France | West TGV (1st section) | 176 km |
| 1990 | France | West TGV (2nd section) | 106 km |
| 1991 | Germany | NSB Wurzburg-Hannover (2nd section) | 237 km |
| | Germany | NSB Mannhein-Stuttgart | 100 km |
| 1992 | Spain | Madrid-Seville | 471 km |
| | Italy | Direttissima (3rd section) | 24 km |
| | France | Lyon By-pass | 38 km |
| | | TOTAL | 1,883 km |

Fig. 5.12   Construction of High-Speed Rail Network in Europe from 1981 to 1992.

Openings through 1996

| 1993 | France | North TGV | 333 km |
|------|--------|-----------|--------|
| 1994 | France/UK | Channel Tunnel | 50 km |
| 1994 | France | S.E. TGV to Valence | 83 km |
|      | France | Link betw. North and SE TGVs | 70 km |
| 1996 | Belgium | Brussels-Lille | 71 km |
|      | France | West TGVs extension | 32 km |
|      | Austria | Wien-Bruck and der Mur | 35 km |
|      |        | Subtotal | 674 km |

Openings expected for 1998-2000

| Germany | (Hannover)-Wolfsburg-Berlin | 158 km |
|---------|------------------------------|--------|
| Germany | Wolfsburg-Braunschweig- | 12 km |
|         | (Hamburg)-Uelzen-Stendal | 15 km |
| Belgium | Brussels-(Aachen) | 84 km |
| France  | S.E. TGV to Marseille & Montpellier | 295 km |
|         | Subtotal | 564 km |
|         | TOTAL for Figure 5.13 | 1,238 km |

**Fig. 5.13  Construction of High-Speed Rail Network in Europe, 1993 to 1998/2000**

## 5.5 CONCLUSION

The considerations above are only a brief account of the transportation planning effort now in progress in the EEC and which could soon extend to other European countries.  In modern economies transportation planning is a necessary and permanent task.  It is even more so in a group of nations that have agreed to unite together to form a multinational economic and political community.  At present, in all member countries, detailed investigations are continuing to test the economic feasibility and possible funding of the various projects outlined.

It is now the task of engineers, but also of economists, politicians, and financiers to elaborate on these proposals and define adequate transport policies designed to implement these projects into a coherent and multimodal system.  No transport system is an end in itself; its basic purpose is to serve the economy, i.e., to respond to the travel desires and transport needs of all citizens within specified economic constraints. Thus, it has to adapt to the new trends in manufacturing and distribution as well as to new transport technologies. If ill-adapted, transport infrastructures may be a hindrance to economic progress and regional development.

It is also clear that in the case of Europe, regional development should not be overlooked. The completion of the single market will induce European firms to establish new plants and create jobs in regions providing the best conditions of accessibility and movement. This will increase competition between European cities and regions; it will also foster tourism throughout Europe, thus introducing people to a larger variety of cultures and facilitating European integration.

Therefore, a primary objective of European transport policies is to ensure continuity of major infrastructures throughout the EEC and, in particular, to provide the links usually missing or insufficient, between national systems. The final infrastructure network should take into account the advantages of new technologies such as high-speed rail, and pay increased attention to energy savings, pollution controls and, in general, the protection of the natural environment. It is hoped that the principles outlined and examples shown for Europe may also inspire new solutions to the transport problems encountered in other regions of the world.

## Notes

1.    The author expresses his thanks to several colleagues:  Jacques Bourdillon, who was in charge of coordinating the transport studies, and Michel Walrave, Secretary General of the International Union of Railways (U.I.C.) who initiated the study of the high-speed rail network in Europe. They have provided me with current infomration on their investigations and on the status of future transport schemes under consideration.

# 6

## Lessons Learnt from Major Projects

### Uwe Kitzinger, C.B.E.
*Harvard Centre for European Studies,*
*and former President, Templeton College, Oxford University*

### 6.1 HISTORY

My story starts at Charles de Gaulle Airport in the winter of 1979-1980, when my old friend from student days, Derek Fraser, then adviser to Swiss Re, told me that the banks and insurance companies were losing large sums on major projects. Things always seemed to go wrong with the projects somewhere. These very big projects overran budgets, overran schedules, or failed to achieve at completion the objectives for which they were undertaken in the first place. The companies wanted to find out why this happened so often. They wanted, if possible, to weed out at an early stage the projects that were bound to fail, and help promote instead mega-projects that could and would actually succeed.

There was evidence that such major projects, whether in aerospace or under ground, whether in telecommunications, transport, or energy generation, in defense systems or in complex organizations, were not just quantitatively, but qualitatively different from minor projects. Indeed, they seemed to have more in common with each other across sectors of industry than with minor projects in their own sector, so that it would make both academic and practical sense to study major projects - regardless of their particular sector - as a class apart.

In early 1981, on the advice of Derek Fraser and Allen Sykes (who also played a sterling role in this), I invited to Oxford a missionary to enlighten the natives. He was one Dr. Frank Davidson, who told us his most successful lectures at MIT had been on "Failure". We asked a variety of people from London to hear him, and that evening enough of them were converted for us to invite further groups to the Oxford Management Centre for several private overnight discussions.

Our guests were originally some two dozen bankers, insurers, contractors, oil, mining and aerospace engineers, accountants and consultants. The purpose was to talk off-the-record in mutual confidence about projects and problems. As our pilot project we organized an overnight discussion of the (then abandoned) Channel Tunnel.

From the mutual discoveries at these first Oxford discussions, four things quickly became apparent. First, even within a single corporation, there was only minimal transmission of the experience of major projects by individual participants. (The senior-level career of few managers spanned the life-cycle of more than one mega-project, and no one was particularly interested in analyzing the causes of failure or indeed those of success. By the time a mega-project succeeds, the responsible leaders are often retired; if it fails, they prefer to forget it as soon as possible.) Second, there was little post-facto contact between the various key participant organizations to compare impressions or derive lessons learned. (The participants in the Channel Tunnel seminar were astonished to learn for the first time from each other the history of that failed project in the round.) Third, there was no coroner to hold an objective inter-disciplinary and inter-industry inquest over the outcome to compare the experience of the different functional participants - governments, owners, financiers, engineers, contractors, lawyers, and the rest. Nor, fourthly, could we in Britain find any analytic historian to compare the problems of different mega-projects in different sectors and different societies to derive general lessons from their successes and failures.

By spring 1982 we were sufficiently confident of the value of our formula of multi-disciplinary, multi-industry, off-the-record sharing of knowledge and experience to constitute ourselves as a professional body with Articles of Association under British law (and in 1987 we were incorporated as a Company Limited by Guarantee). We called ourselves the Major Projects Association - referred to as the MPA for short. Its aim, in the words of its Articles is: "...to enhance the ability of its members to initiate, assess, secure and accomplish successful major projects".

## 6.2 WHAT CONSTITUTES A MAJOR PROJECT?

Few of us have difficulty recognizing a project: we immediately think of an undertaking that pulls together very diverse resources to achieve a complex objective defined usually in terms of technical performance, budget, and time and quality constraints, and with a sequence of activities from conception and formulation through feasibility studies, design and contract negotiation to implementation and (where appropriate) handover and running support. At least since Gaddis' eight-page article on the subject in the *Harvard Business Review* in 1959, Project Management has been recognized as a subject that ought to be researched and taught in business and engineering schools. But, until recently, it was seen and taught as a nitty-gritty set of internal organizing techniques applied to multifarious complex tasks - critical path scheduling, matrix systems of organization, quality assurance and the like rather than as an inter-disciplinary, inter-sectoral, geopolitically strategic view of what we now refer to as major projects.

As to what constitutes a "major" project, I remember when Frank Davidson and I toured Japan in 1981 at the invitation of Professor Manabu Nakagawa, we spent some time debating how to define a "major" as distinct from a "minor" project, macro-engineering as against micro-engineering: whether sheer cost, or size, or number of participating organizations, whether political or environmental impact, whether technological innovation or substantial irreversibility (or some combination of such criteria) should delimit the concept.

In the MPA, we are empirical rather than Cartesian. For practical purposes we now treat as "major" "any collaborative or capital project requiring knowledge, skills, capital, or other resources that go beyond those normally available to the key participants in it." No wonder, then, that working at the limits of previous human achievement, major projects (as we define them) are by their nature challenging, stretching, often frustrating, and usually fun.

Let me list some of the engineering projects that the Association has been studying to give you an idea of the range of evidence we consider relevant:
- the Thames Barrier to save London from inundation;
- man-made islands;
- the Sizewell B nuclear project (with a design staff of 1,500 people and a seven-year public planning inquiry)
- the Tornado military aircraft, Concorde, and the European Space Agency's successful project "Giotto", to intercept Haley's comet on 13th March 1986 as internationally collaborative projects;

- the Joint European Torus nuclear project (in which the project director
  had to report to a governing body of political representatives of eight
  European states).

This study of past projects reflects the inevitable need to profit from past
experience.    Naturally our members are also interested in future needs and
opportunities, and so some of our seminars have concentrated on opportunities for
future major projects in India, China, Indonesia, the Middle East, and in the
former Soviet Union.    But on the whole we look to the nearer opportunities
rather more than does the American Society for Macro-Engineering; and in most
of our activities we are perhaps not as *avant-garde*, as visionary, as others may
be.  But our technical papers do include "Aerospace 2000", "Surface Transport
- The Next 20 Years" and even "Information Technology - The Next 35 Years".

Please note, however, at the same time that we did not want to use the term
"macro-engineering".  That was not simply out of a rather British phobia of words
like "macro", but also because we could foresee that some of the projects we
might study hardly qualified as "engineering" in the strict sense of the term.  We
would have regarded, for example, the Normandy landings of D-day 1944 or
"Desert Shield" and "Desert Storm" in 1990-1991 as major international projects
- and from the list of our case studies you will appreciate why.  It includes what
some people call - perhaps misleadingly - "soft" projects such as:

- COP (the computerization of the British Pay-As-You-Earn tax system);
- the disastrous attempt to set up TAURUS - the Transfer and Automated
  Registration of Uncertified Stock (abandoned in 1993 after five years'
  work); and
- the much more successful "Project Sovereign" completely revamping the
  organization of British Telecom and reducing twelve management
  levels to six.

In the case of more general problems, our events have dealt with, among
other not necessarily engineering subjects, "Hazardous Waste", "The Environment
and Global Warming", "Road Privatization" and "Cleaning up the Mediterranean".

In fact the lessons we draw from major projects do not usually refer to
technical matters except in one respect: the profound effects that technical
uncertainty and technical advance can have on the progress through time of the
feasibility, coherence, and indeed even on the ultimate objectives of the whole
project itself.  When we step back from individual case studies, some of our
eighty or so seminars have focused on broader aspects of major projects such as
"Unresolved problems of risk," "Current changes in contracting practice",
"Relationships and motivation in international projects", "Causes and
consequences of delay", "Criteria of viability", "The balance of experience and
uncertainty", International negotiation", "The environmental, health and safety

factor in major projects", "Political risk", "The peace dividend" and "Asian idioms of Capitalism".

Our basic premise - that major projects, in whatever sector, were not just quantitatively, but qualitatively different from minor projects and had more in common with each other than with minor projects in the same sector - has been amply confirmed over the past fifteen years. Over and above technical problems and the usual problems of project organization, they pose a series of external dilemmas: dilemmas of long-term mega-finance and mega-insurance; dilemmas of contract law and contract practice; of cross-industry and particularly of multi-cultural managerial collaboration; environmental dilemmas; issues of community relations; and dilemmas deriving from politics (both national and international). These, rather than mistakes in engineering technology, are the challenges that are more likely to pose the real problems - and risk plunging a project into failure.

So our philosophy boils down to the conviction that as war is too important to be left to the generals, so macro-engineering is too important to be left to the engineers. If the most salient characteristic of major projects is the complexity of necessary interactions between the participants, then the real problems are not technical problems, but problems of relationships. We are talking here of human relationships, financial and organizational relationships, and political relationships. These depend on attitudes and subconscious assumptions, habits and stereotypes, contacts and communications, negotiations and contracts; and they express themselves not least in the manner in which the key stakeholders in such projects - the ultimate owners, the project management teams, and the often hundreds of suppliers of the vast and multifarious resources entailed - analyze and allocate (whether by contract, or by implication, or by default) the total risks and rewards of failure and success, and above all, of partial failure and partial success -- risks that are a combination of the technical, the financial, the economic, the environmental and the political. In other words, macro-engineering involves not only technical engineering but financial engineering, environmental engineering, social engineering and political engineering. It was, therefore, in our view, entirely appropriate and consonant with our perspective and approach that the host organization should be not a School of Engineering or Technology but a Centre of Management Studies - what in 1984 became Templeton College and, with its Royal Charter granted last summer, is now the 37th College of Oxford University.

## 6.3 THE ASSOCIATION

So much for what we ambitiously consider as our bailiwick. Now about our organization. After the first five years, I passed the chair on to a prominent British industrialist, the late Sir Alistair Frame, and the third Chairman is a retired head of the British Foreign Office, Sir Michael Palliser. The Board includes, for example, Mark Stephens, the Executive Director of the Berne Union and Tony Ridley, President of the Institution of Civil Engineers. After several men from industry, the present part-time Executive Director is Christopher Benjamin, a former senior civil servant in the British Department of Trade and Industry (which he had represented on our Board from the beginning) who now also works for a major Japanese trading corporation.

It is a considerable - indeed, a key - difference compared with the American Society for Macro-Engineering that the members of the Major Projects Association are not individuals. The members are organizations - most of them corporations - able to commit themselves long term through all the inevitable changes in personnel, and prepared to field expert teams for the work of the Association; they are also of course able to pay a useful membership fee (now about $6,500 plus value-added tax). To become members they must undertake to respect the confidentiality of MPA proceedings, to share with their fellow members a distillation of their experience regarding major projects, to participate from time to time in preparing a detailed case study or topic presentation, and to be represented regularly at senior level at our seminars and workshops and our annual General Meeting. It follows that they do not always have to be well-known names or very large corporations: what matters is that they are able and willing to contribute significantly to the overall running and success of the Association's activities.

We are a private, not a public association. The Templeton Seminars must not get too big or they will lose effectiveness. Given limited numbers, we have to maintain a balance between different industries and services and between the different functions performed in initiating, assessing, financing, insuring, completing and managing a major project. Membership is therefore by invitation, new members being elected by the Board. I can remember no instance of a member hesitating about a competitor anxious to join. The members regard this not as a competitive, but as essentially a collaborative professional endeavor.

Originally some members thought of the MPA as a British organization, helping British firms win contracts for major projects abroad against foreign competition. But we have never seen that as an exclusive aim or one incompatible with drawing on the widest possible experience to enhance our professional competence. As a result, a number of multinational or even predominantly non-British firms have joined.  How far we should now go to extend full or corresponding membership (the difference between the two classes of membership is progressively eroding) on a worldwide scale (and what that might mean for numbers) is a matter being re-considered at the moment.

Our membership quickly rose to over fifty, and it may be worth listing a few of our members here.  Those beginning with the letter B include Balfour Beatty, Bechtel, Bovis, the British Airports Authority, British Gas, British Railways and British Telecom, and darting round the rest of the alphabet we find, among the better-known names, Arthur Anderson, Coopers & Lybrand, Eurotunnel, IBM, McKinsey and Rolls-Royce.  Among non-profit organizations we have (as the MPA's host) Templeton College but now also its rival Henley Management College, the Civil Aviation Authority, and among other British government departments, the Ministry of Defense.  The European Investment Bank figures as a corresponding member together with Schiphol Airport, Crédit Agricole, and the Japanese construction firm Taisei Corporation.

So what does a member organization get out of belonging?  Above all, it is the right to be represented at senior level in the private seminars at Templeton College, usually held from Thursday before dinner until Friday after lunch.  There cards are put on the table with sometimes astonishing frankness as to mistakes and bluntness as to persons.  The proceedings are recorded and edited - every participant has the right to strike out anything he said which he regards as too delicate for even a confidential written record - and then printed in sufficient numbered copies to go under seal to each of the full member organizations on the strict understanding that it is only for the internal use of member companies (or the other speakers at that seminar itself).

At the same time we should not underestimate the value of the very natural informal cross-disciplinary and cross-industry encounters in a quiet place like the bar of an Oxford College and the comradeships formed at what we term the "constructive boozing" which often goes on well after the Thursday evening session has adjourned.

Some years ago we started to supplement the private seminars at Oxford with lectures - often public lectures - in London, particularly with speakers from abroad.  One of our first was on the subject of the impressive success of the United States' Apollo project.  We also organize field trips to places like the Thames Barrier - this in conjunction with an overnight Templeton College seminar on the same project.  More recently we have started a series of London lunches for senior executives from firms working in the field who may not actually be members.  Our annual conferences are focussed on themes: the fifth, in 1987, studied "Planning and Enquiry Processes"; at the twelfth, held in October 1994, it was "Making Major Projects Happen" with lectures on the role of government, the support of the European Union and its agencies, the role of the private capital markets, issues surrounding contracts and their negotiation and the management of environmental factors. The Autumn 1995 annual conference concentrated on "The Changing World of Risks".

For the past thirteen years we have also run residential training courses now entitled "The Challenge of Major Projects" to pass on to managers in mid-career the lessons we are learning from the Association's activities. These courses now figure twice a year, in May and in November, on the Templeton College calendar: they are charged at some $4,000 per participant for the five-day week, and focus particularly on the factors which traditional project management courses hardly touch: the environment, politics, conflicts of interest, legal and contractual organization, the distinction between the promotion, ownership, and provision of major projects, the definition of objectives and the assessment of risks.

All this is run by the Executive Director together with an administrator and an organizer of events. Their offices are in Templeton College itself. The annual budget, financed above all by the membership fees but also by the training courses and miscellaneous other earnings, is steady in the region of $400,000 to $450,000.

## 6.4 SOME PERTINENT QUESTIONS

One may well ask how far a transposition of this formula for confidential inter-corporate and cross-industry meetings would be possible in the United States anti-trust context. Some American lawyers are of the view that beyond a certain level in the corporate hierarchies, it could be dangerous for senior executives from different companies to attend such meetings. In Britain we have never had the slightest problem on that score: in fact, Her Majesty's Government itself came in at a very early stage with a triple membership subscription which is used at various times by the Foreign Office, the Department of Trade and Industry, the Department of Transport and others.

But behind that legal question there no doubt lurk broader issues. What, in particular, is a political scientist and head of an academic institution doing helping set up, hosting, and being founding Chairman of such an inter-industry body? And what, in general, can be the social utility of such an inter-disciplinary Association?

To me, the answers to both these questions are perfectly clear. First of all, there is the academic synergy. We can have no up-to-date and relevant teaching without up-to-date and relevant research. As more and more managers are drawn into major project work, more and more of them need to be made aware of the peculiar problems they pose. And in any case the study of limiting cases - whether in complexity or in size - is an essential part of the study of the management of a whole gamut of projects. Indeed, we found that the analysis of failure and success not only highlights key factors that distinguish major from minor projects, but also casts interesting light on the problems of managing minor projects as well. To take but one example: the problems of major projects make explicit the types and degrees of risk absorbed implicitly by participants in more usual lesser projects who can spread those risks over the many other irons they have in the fire at the same time. So both teaching and research in major projects are matters of urgency for any College of Management Studies. Templeton College has become noted in this field, particularly through the series of short courses for mid-career managers organized and run there by the Major Projects Association.

Nor do I see any problem for collaboration of this kind between academic, business, and government organizations arising from the distinction between confidential and public information. As professionals, economists and political scientists are used to handling confidential information as evidence, although as academics we owe our conclusions to the public. Similarly in the sphere of industry and for-profit services, let alone departments of government involved in the assessment, promotion, negotiation and support of major projects: fully detailed information can be obtained only on condition of confidentiality. But the general conclusions and maxims of best practice derived from those data must be passed on and published for the benefit of industry, government, and society as a whole. That is precisely what the Major Projects Association, as a professional body, does through its series of publications - its research papers, reports, pamphlets and books that are publicly available and through its professional training programs. Our motto, paraphrasing Woodrow Wilson, can be "open lessons, confidentially arrived at".

In addition to these intellectual benefits to research and teaching, both Templeton College and the MPA derive quite practical benefits from their symbiosis. These are benefits in terms of sharing links to a variety of corporations, benefits in terms of public recognition, and even to a minor extent some mutual financial benefits as host and guest, landlord and tenant. The MPA in fact takes its place alongside various other "parallel" activities begun at the College in the early 1980s, notably the Oxford Economic Forecasting unit, which does a good deal of client work but whose general forecasts are regularly published; the Oxford Institute of Retail Management, which is consulted by retailers from all over Europe; and the Oxford Institute of Information Management, which similarly combines research and teaching as part of the

College's programs. All these have greatly enriched the intellectual substance and the practical relevance of the College.

But coming to the broader question of the public interest first, the enormous demographic, resource, economic and environmental problems of this planet will not be solved without commensurate projects to tackle them. To prevent further despoliation of the planet, to ensure a decent quality of life for many human societies, to pep up world economic and social progress and reduce unemployment, major projects are more crucial than ever before. Small may be beautiful, but without macro-protection and macro-infrastructure small-scale units are vulnerable and often even impossible to develop. Even rural smallholders may need big irrigation schemes; family homes in vast Third World urban centers need water and sewage; one-man workshops need public infrastructure for energy and transport. Major projects are likely to become a prerequisite of civilized human survival. These global issues must be a concern for all who care for the future.

Second, there are the specifically international aspects: the disparities between the West, the former Soviet sphere, and much of what used to be lumped together as the Third World will not be mitigated without very major undertakings of flood control and of beating back desertification, not to mention the development of cleaner sources of energy such as solar, wave and wind power, of transport, information and communication projects, and of mass education. Moreover, undertaking together internationally designed, financed, and managed projects could and should act as a peace-building cement between too often disunited nations, religions and ethnic groups. These two international considerations - one on the demand side and one on the supply side - are also part of the thinking behind the European Round Table of industrialists and its European Centre for Infrastructure Studies in Rotterdam (with which the MPA now has cross-membership) and of the Japanese-based Global Infrastructure Fund. Both these international aspects are again issues of world citizenship.

## 6.5 SOME LESSONS

Now let us turn to some of the conclusions to which these fifteen years of collaborative study have come. (I do not say at which we have arrived, because in academic matters it is better to continue to travel than to believe that you have ever finally arrived.)

The first lesson is perhaps the most obvious - yet how often is it really followed? A precise definition of the project - its objectives and its means, its priorities and its limits - has to be understood, agreed, and internalized by all the participants.  To go back to our first case study: was the aim of the Channel Tunnel to dig a long continuous hole in the ground?  If so we have now succeeded.  Was it to provide a substantially faster route from London to the nearer continental capitals?  If so, we are still far from success.  Whether a dam is to prevent flooding, to produce energy, or to demonstrate national prestige and consolidate some dictator's political power needs to be clarified at the outset, and the means then need to be calibrated to the priorities agreed.

One of the earlier broad evaluations of our work was Derek Fraser's pamphlet for potential owners (particularly in the Third World), now available in revised form under the title *Initiating Major Projects*.[†]  I commend it to you as an excellent cautionary introduction compact enough to be read by a senior minister on a single flight.  One of the points he makes is that a serious feasibility study can last for years and eat up some 3-5% of total project cost - often a prohibitive sum.  That is why he insists on what he calls a "credibility study" long before a feasibility study is commissioned.  And let us always remember that it is the first 10% or so spent in the conception and strategic design of a project that matters far more to its success or failure than the other 90% spent in its execution.

In 1987 the MPA produced a book of over 300 pages, *The Anatomy of Major Projects* by Peter Morris and George Hough which tested some 22 hypotheses against a series of our case studies.  The authors come up with a list of eighty factors for project success (see Exhibit 6-1).  Some of these maxims of best practice look obvious enough to be managerial "motherhood and apple pie" - but then the case studies remind us starkly how often they can be forgotten, neglected, or impossible to achieve in practice.

The best easily available summary of where we think we stand now is to be found in the collective volume *Beyond 2000 - A Source Book for Major Projects* revised in 1994.  Even a quick inspection will reveal how far we have broadened our purview.  In the opening section we focus on the needs entailed by world population trends, urbanization, and the deterioration of the environment with the depletion of the ozone layer and global warming, and discuss the likely impact of religious fundamentalism and crises of ethnic identity on the now urgently necessary global cooperation.

The MPA pulls no punches on some of the issues involved, and you will get a flavor of it from selected quotations.  The first lesson reads: "Major projects intersect sharply with environmental issues.  No project is an island..."  We praise the U.S. for its environmental assessment requirements and look forward, under the European Environmental Assessment Directive 85/337 to "best practical

---

[†]   For a bibliography of the MPA's own publications, see Exhibit 2.2.

means" being replaced by "best available techniques not entailing excessive costs". Not every member firm would agree wholeheartedly with our conclusion that: "Nongovernmental organizations have a crucial role to play ...a role of enlightened responsibility".

We castigate the British - dare I say the Anglo-Saxon - tradition of adversarial and short-term relationships, and the fashionable "invisible hand" doctrine that competition is the ultimate test of efficiency and the road to optimal outcomes even in gigantic endeavors. We deplore the deep-seated British dislike of comprehensive strategic thinking, as witnessed in stop-go economic policies, out-of-date accounting systems, the lack of investment in new technology and new plant, the gross failures in education and training. Add for further spice a swipe or two at individualism and entrenched conservatism in many industries and across much of the population.

Governments, for both long-term ideological and short-term financial reasons, have become too chary of supporting projects that could attack major problems, provide employment, and stimulate economic growth. As attitudes to risk have hardened, the separation of private and public finance is unlikely to be maintained, and complex multi-sourcing packages of private, national, and international finance with intricate overlays of private and government guarantees and insurance are likely to become more common. Recent MPA events have looked particularly at the problems these are liable to pose, including, on a European scale, the Trans-European Transport Network projects.

Much discussion is now focused on new financial relationships between the private and the public sector, like "BOOT" (Build, Own, Operate, Transfer to public ownership); there are new formulae of relationships between what used to be the distinct and almost adversarial categories of owners and contractors - "alliance contracting" is another formula that has already proved itself in at least one case as a great saving of cost. Managers of major projects are learning to turn to their advantage the way different contract provisions affect the allocation of risk, how this in turn affects the motivation of the contributors to the project, and how contracts influence the flow of information between organizations within a project. There is indeed a great deal more intellectual work to be done to invent and to hone innovative relationships.

Then again we need to be sensitive to the added dimensions of risk posed in the case of major projects in Information Technology. Apart from a number of case studies in major Information Technology projects, two of our earlier events dealt with "Safety critical systems and software management" and "Artificial intelligence in engineering and the management of major projects - hope or reality?" Two recent seminars, one on "Integrated Systems" and one on "Networks" have sought to face the IT snake-oil salesmen and asked if the new technology really gets the essential information clearly to the decision-makers rather than confusing issues and generally introducing its own temptations and complications. When one has to learn from the past experience of major projects at the same time as learning the potential and the limits of new technology, then a less hazardous alternative to major projects may be incremental innovation. Either way, the risks in information technology need to be profiled with particular precision.

In the future, the style of major project managers will be - as it has to be - increasingly different from that of managers elsewhere. Power over resources may have to be given to one person who will use it to avoid as well as to manage problems. Systems and organizations will have to be invented afresh and perfected project by project. Training in project management has to change: it has to be given to people early in their project careers and concentrate on illustrating the general principles and the attitudes required in the majority of such projects. Maybe once they can show the requisite experience as well as the theoretical grasp, project managers should indeed be certified - and there are now moves afoot for such certification in Britain, and so are moves towards a European standard.

Overall, I believe we can be moderately optimistic. Major projects as a distinct field of activity, of research and of teaching have come a long way over the past decade or two. There are no dramatic new solutions, and there are of course - almost by the definition of major projects - no standard answers. But there is far greater awareness of the interdependence between a technical project and its social, environmental, and political context. On the whole, the risks of cost and time overruns and of poor performance can now be reduced to tolerable levels. If the Major Projects Association has made a certain contribution toward that end, then Frank Davidson's inspiration and the efforts we put into setting it up fifteen years ago will have been amply worthwhile.

But though our studies and, to some extent, our practice have progressed, neither has advanced anything like enough. There is an imperative need for continued endeavors - yours in the Society for Macro-Engineering, ours in the MPA, and those of our sister organizations in Japan, France, Spain, Canada, Scandinavia, Holland and elsewhere. We still have a real job to do - and we must do it together.

## Notes

For their comments on an earlier draft, the author wishes to thank Chris Benjamin, Janet Caristo-Verrill, Derek Fraser and Jenny Kitzinger.

## Exhibit 6.1
### Factors For Success

*Project Definition*
    Communicate clearly
    Phase as appropriate
    Identify, assess and develop sub-objectives clearly
    Relate objectives to participants
    Do not force clarity until appropriate
    Beware of progressive change
    Avoid too early a commitment

*Planning, design and technology management*
    Attend to broader, systems aspects of projects
    Relate to phasing, logistics, geophysical uncertainties, and the design and
        production relation
    Have back-up strategies for high risk areas
    Develop the accuracy of estimates to an extent consistent with the
        uncertainties present, e.g. technology, methods
    Avoid concurrency (see below)
    Test design adequately before final project commitment is made
    Recognize the extent to which R&D is completed will affect accuracy of
        estimate
    Use flexible design philosophies
    Recognize that good design management is essential, especially where there
        is technical uncertainty or complexity
    Recognize that interface management is important where there are significant
        interdependencies
    "Freeze" design once agreed
    Beware of switching design authority during different phases of project
    Pay attention to detail since mistakes can prove costly
    Encourage replication where appropriate

*Political Social factors*
    Ensure effective sponsorship
    Recognize fiscal, safety, employment, etc., constraints
    Ensure support for such management actions as may be necessary
    Constrain nationalistic aspirations on international projects
    Manage community factors effectively

**Exhibit 6.1**
**Factors For Success**
(continued)

### *Schedule duration*

Recognize the major impact that output, price, regulation, technical developments, government or corporate changes can have on definition of success

Phase projects where possible to avoid unnecessary over-commitment

### *Schedule urgency*

Avoid rushing

Note possible disruptive effect on work sequencing

Beware of impact on full discussion by all parties

Beware when urgency and technical uncertainty go together (concurrency)

### *Finance*

Undertake full financial analysis of all project risks: budget validity, political support, owner's commitment, etc., including inflation, and possible currency variations

Be cautious over availability of funds

Be prepared to stop funding where necessary

Seek sponsors interested in success of project *per se,* not just a good return

Beware of exchange rate movements

Check definition of project success if business base of project changes

### *Legal agreements*

Ensure break clauses are adequate

(Beware of 50-50 partnerships)

(Beware of mixed public-private funding)

Seek commitment to making contract work

### *Contracting*

Consider whether more innovative contractual arrangements may not be required

Consider incentive contracts valuable where it is difficult to get competition, though beware of too high a level of technical uncertainty

Ensure contractors are sufficiently experienced to perform the work

Consider the extent to which competitive bidding is appropriate

Beware of the same organization acting as contractor and owner

## Exhibit 6.1
## Factors for Success
(continued)

*Contracting*  (continued)
>    Provide adequate bid preparation time
>    Beware of the cheapest bid
>    (Beware of having to manage a large number of contracts)
>    Define contractor's responsibilities clearly
>    Make contractors financially responsible for their performance as far as
>        possible
>    Beware of contract forms which unfairly penalize contractor, particularly for
>        factors outside his control
>    (Beware of mixing firm price and reimbursable forms)
>    Question the threat of liquidated damages
>    Appraise carefully whether interference by the owner in the execution of a
>        contract is justified

*Project Implementation*
>    Seek appropriate client, parent company and senior management attitudes and
>        support
>    Control all those aspects of project which can affect the chances of success
>    Recognize the magnitude of task and organize appropriately
>    Obtain clear client guidance
>    Foster good client-contractor relations
>    Integrate the project teams' perspectives with the project aims during start-up
>    Assess risks adequately
>    Develop good planning, clear schedules, adequate back-up strategies
>    Exercise firm, effective management from the outset
>    Recognize the importance of effective, schedule-conscious decision making
>    Provide clear and comprehensible project organization appropriate to the size,
>        urgency and complexity of the project
>    There should be one person, or group, in overall charge having strong overall
>        authority
>    Ensure effective leadership
>    Strive for a well motivated, experienced team
>    Develop appropriate controls, highly visible, simple and "friendly"
>    Check definition of success, where changes are allowed
>    Ensure resources are adequate, properly planned and flexibly employed
>    (Consider use of site labor agreements)

**Exhibit 6.1**
**Factors for Success**
(continued)


*Project Implementation* (continued)
    Ensure labor practices are consistent among and between contractors
    Give full recognition to quality assurance and auditing
    Recognize that good communications are vital

*Human factors*
    Ensure top management support
    Recognize and demonstrate the importance of effective leadership
    Seek competent personnel
    Ensure communications are effective
    Consider which power style is appropriate
    Recognize that people are human and less than perfect


Source: Extracted from P. Morris and G. Hough, *The Anatomy of Major Projects*. Wiley, 1987, pp. 265-66.

## Exhibit 6.2
## MPA Publications

"Adapting to Future Needs and Markets in Construction," Technical Paper No. 10, August 1991.

*Looks at what types of major construction projects are likely to occur in the years up to 2010, where significant activity will take place, and how projects may be financed and executed.*

"Aerospace 2000," Technical Paper No. 14, February 1993.

*Within the worldwide scope, there is a concentration on UK interest in civil aviation and space systems in a European context and on the prospects for major projects.*

Allen, J., "Man-Made Islands," Technical Paper No. 4, May 1987.

*A review of developments to date, and the challenges and opportunities of these structures.*

"Beyond 2000: A Source Book for Major Projects. The context, prospects and management of major projects into the next century."

*The Source Book is an indispensable guide to developments in the next generation of major projects in the key areas of energy, aerospace, transport, and construction.*

Cloot, P., "BBC Producer Choice: A Case Study," Briefing Paper No. 16, May 1994.

*An independent assessment of the BBC's internal market system.*

Fraser, D., "Initiating Major Projects," May 1994.

*The problems and solutions of initiating major international projects.*

"Information Technology: The Next 35 Years," Technical Paper No. 12, July 1992.

*This paper concentrates on the evolution of IT itself and the nature of the opportunities suggested by the technological trends identified.*

## Exhibit 6.2
## MPA Publications
(continued)


"Major Projects and the Environment," Technical Paper No. 8, June 1989.
*An immensely valuable collection of papers by project practitioners, environmentalists, and others offering both an objective evaluation and occasionally an impassioned account of the interaction between major projects and their environment.*

Morris, P.W.G., "Issues Raised in Seminars of the Major Projects Association, December 1981-June 1984," Technical Paper No. 1, July 1985.
*An account of the principal points raised in the MPA's early years.*

Morris, P.W.G., "Managing Project Interfaces - Key Points for Project," Technical Paper No. 7, June 1989.
*An analysis of the organizational framework underlying the project manager's job as he steers his project through its life cycle.*

O'Riordan, T., "Major Projects and the Environmental Movement," Technical Paper No. 5, May 1988.
*An important essay describing the nature of the environmental movement and how the impact of major projects on the environment can be better managed.*

"Project Opportunities in the Energy Industries," Technical Paper No. 11, February 1992.
*The driving forces of the huge predicted growth in the energy industries are detailed in this paper, together with where and in what sectors this growth is likely to occur.*

Stringer, J., "Planning and Inquiry Processes," Technical Paper No. 6, September 1988.
*A unique survey of how the early planning and consents phase of a project should be managed for the benefit of all concerned.*

**Exhibit 6.2**
**MPA Publications**
(continued)


Stringer, J., "Gaining Consent - Guidelines for Managing the Planning and Inquiry Stage of Project Development," Technical Paper No. 9, August 1991.
*Step-by-step guidelines to the serious hurdle of obtaining planning permission, facing a public inquiry, or getting a Bill through Parliament for a major project.*

"Surface Transport: The Next 20 Years," Technical Paper No. 13, July 1992.
*This paper describes the present position of surface transport and then goes on to detail the likely developments.*

# 7

# Guided Transportation Systems: Low-Impact, High-Volume, Fail-Safe Travel

**David Gordon Wilson**
*Professor of Mechanical Engineering*
*Massachusetts Institute of Technology*

## 7.1 INTRODUCTION

Three types of wholly or partly new transportation systems are described. One is based on human-powered vehicles (e.g., "cycles"); one on buses; and one is a replacement for conventional rapid transit called Personal Rapid Transit (PRT), currently being planned for an area of Chicago.

These systems are of interest now because of problems in our present ground and air transportation systems: while they serve some purposes and some groups of people very well, they serve many other purposes and many other people quite poorly.

The reasons for this state of transportation to persist at the end of the 20th century are that overt and concealed subsidies for some forms of transportation have become entrenched in our national lives to such an extent that we regard them as normal.

The introduction of new transportation systems needs either the imposition of a free market in transportation, or the intervention of strong, wise government, or of an equally wise and strong - and very rich - private benefactor or entrepreneur. The author's preference is for the free market. That would, however, involve measures to recover external costs that will seem suspiciously similar to taxes. At present, governments - strong, but not always wise - are taking the lead.

## 7.2.    THE TRANSPORTATION PROBLEM:
## HOW DID WE GET THIS WAY?

Americans are in love with their automobiles. They are not alone. Citizens of almost every other nation on earth feel likewise. Even China is planning to reduce its bicycle population in favor of giving more space for motorized transportation. And every year, automobiles get better. They pollute less, in general they use less fuel, they are safer, they have incredible built-in gadgets - motorized adjustable lumbar supports, satellite route finders, and remote-controlled windows, for example - that we have in no other area of our lives. An automobile can take a family in superb comfort in almost any weather to any part of the country. They are seen performing these wonderful feats in mouth-watering advertisements on television and in magazines.

In these advertisements the car being advertised is almost always entirely alone on a beautiful highway or in front of a stately home. In these circumstances, which the advertisers would want us to believe are the norm, the car brings benefits to its users and disbenefits to none.

In reality, of course, the great majority of automobiles are used in the hundreds of thousands, sometimes in the millions, in towns where they bring great disbenefits, mainly in the form of delays, to other automobile users and far greater problems to nonusers. Congestion in cities is so great that in many the average speed is lower than in the days of the horse and cab. Smog is damaging to health and even life-threatening. Global warming is blamed to a large extent on automobile production and use. Automobiles have made cities so congested and dirty that anyone who can afford it uses an automobile to travel to and from a cleaner, more prosperous suburb, thereby contributing much less to the city's now-swollen budget. Thus, cities have become huge pockets of poverty and crime, abandoned every evening by most of the people who work there. Crime is itself enabled by the automobile, either in its commission, as in drive-by shootings, or in the ease with which criminals can escape and merge among other faceless people encased in their steel and smoked-glass chariots.

### 7.2.1    Some Recent Transportation History

At the time the automobile was invented, railroads had been developing apace for more than sixty years - a steam train exceeded 100 mph within a year or two of the appearance of the first motor car. The railroad transformed life for rich and poor, because people who could afford them could travel long distances in great speed and relative comfort (compared with stagecoaches), and because fast low-cost transportation of freight could bring better and/or lower-price goods and could take the products of the workshops and farms to wider and higher-priced markets.

The pedaled bicycle had been evolving for more than thirty years, reaching phenomenal popularity among people who could afford it. Women's liberation traces its origins to the freedom that the bicycle bestowed on women generally: they could not be reliably chaperoned (controlled) on a one-person bicycle. The League of American Wheelmen started the Good Roads Movement to make bicycling much more enjoyable and useful, and that also unwittingly prepared the way for the "horseless carriage" to have an easy introduction.

Street railways, horse-drawn at first, had also been used for nearly thirty years, and underground railroads (subways) were extending fast in London and later in Boston and New York.

Neighbors of the railroad, including cattle farmers, were not universally pleased about its arrival, and teamsters were among those who vigorously opposed bicycles because they frightened their horses. But on the whole the disbenefits brought by these two revolutionary forms of transportation were relatively small and not widespread. And both railroad operators and bicyclists had their lobbyists to try to bring about favorable conditions for their users.

When the automobile was first introduced, it was seen as a threat by many, and the lobbyists of the then-prevailing systems went into action. In Britain a law was passed requiring a motor vehicle to be preceded by a man carrying a red flag. That law was soon overturned, however, and motor vehicle use increased throughout the century except in times of war.

I have memories of my father driving the family to seaside vacations in the 1930s, at which time the number of vehicles on British roads was only a few percent of its present number. Yet, the congestion and long waits seemed insufferable. It must have seemed easy to join in the demand to build bypasses around towns, to widen roads, and eventually to build superhighways. Probably most of the construction funding came from fuel taxes. However, little or no funding went -- or goes presently -- to compensating the many who were and are disadvantaged by new roads: the people who must move to new neighborhoods; the people who can no longer visit their friends across the road; the people who are harmed by air pollution; those who are injured in motor-vehicle accidents and the relatives of those who are killed (50 million deaths in the first century of the automobile.)[1] Then there are even less tangible costs: the small businesses that close when superstores move into the area, dependent solely on shopping by automobile; the urban sprawl that changes the whole character of community life; global warming, and so forth. Motor vehicle-related taxes today do not even pay for road and bridge maintenance (see Figure 7.1). They certainly do not pay for that proportion of our armed forces that could be regarded as needed to defend oil supplies.

Not all readers will accept that these supposed disadvantages and costs are properly named and assigned (for instance, many people prefer shopping malls and supermarkets rather than small-town shops). Another example is the practice by many businesses of providing free parking for employees, which is actually a subsidy of one form of transportation over another. The cost is considerable, but is regarded as part of the cost of doing business. Recently, the Massachusetts Institute of Technology provided free parking for its employees until it was calculated that the cost of providing one parking place averaged out at well over $2,000 per year. The employees who walk or take public transportation or who bicycle are given no such subsidy. Driving a car then becomes a relative bargain. People decide to own and drive a car when, if they had to pay market rates for parking, they might well use other means for commuting, and they might live much closer to their place of work. The community local to the work site has to hire more

---

**DRIVING SUBSIDIES**

"Contrary to popular belief, drivers do not pay their own way through user fees. In the U.S. gasoline taxes and other user fees account for roughly 60% of federal, state, and local spending on highways. The remainder, $29 billion in 1989, comes from general funds, property taxes, and other sources. Another cost, "free" parking, has an estimated value of $85 billion per year. Additional expenses not covered by drivers, such as for police and emergency services, traffic management, and routine street maintenance, represent some $68 billion annually. When harder-to-quantify costs such as air pollution, traffic congestion, and road accidents are figured in, the total subsidy to drivers in the U.S. soars to an estimated $300 billion a year."

(Marcia Lowe in "State of the World, 1993", quoting J.J. MacKenzie et al.)

---

police and traffic wardens and put up more traffic signs and signals to cope with the additional traffic. Roads become more congested and hazardous and people think carefully before letting their children walk or bicycle to school. Instead, children may be driven, or the school system buys school buses and hires bus drivers, adding to the congestion, the pollution, the hazards, and the cost of living generally.  One subsidy thus begets a spiral of costs that are of enormous wide-ranging impact.

My favorite analogy to the present transportation picture is one in which the government distributes free ice cream[2] (see Figure 7.2).  This may be regarded as a proof by induction.

---

### FREE ICE CREAM FOR ALL!

The results of such a policy:

1. The government would become wildly popular.

2. Consumption of ice cream would rise dramatically.

3. Government would pay large sums to farmer, ice cream producers, distributors and retailers.  Employment in these industries would increase dramatically.  A huge lobby in support of these excellent policies would arise.

4. Consumption of "bread and broccoli" would decrease greatly.  Producers of all "alternative foods" would ask for government subsidies, which would be grudgingly given.

5. Obesity and ill health would spread.  Government would institute massive research programs to discover the causes.

6. People would start using the free ice cream to make plastics, feed pigs, and cool their houses.  Water pollution would become serious as sewage plants became overloaded.  Government would set up a special branch of the police with draconian powers to search and seize.

7. The huge increase in consumption, legal and illegal, of ice cream would cause shortages to develop.  Lawmakers would propose rationing and target reductions.

8. Taxes to pay for all these programs would become burdensome.  Economists would earn Nobel Prizes for theories typing market swings, general low agricultural employment, and high taxation to long-wave theories.

Figure 7.2.  The Free Ice Cream Analogy

---

## 7.3 HOW SHOULD NEW SYSTEMS BE INTRODUCED?

The general political approach to this imbalance has been to subsidize competing modes (the broccoli farmers in the analogy). Buses, rapid-transit rail systems, railroads and air transportation systems receive massive federal subsidies. Even bicyclists feel justified in asking for subsidies to try to balance the inequities. The method by which the new forms of transportation systems mentioned below should be introduced, in the minds of the overwhelming majority of the population, is that the government should give the developers of a system a contract. Typical government capital costs for the construction of traditional rail rapid transit, even along existing corridors, have been of the order of $200 million per mile. After the taxpayer funds this huge expenditure, he/she must also pay large annual operating subsidies thereafter.

Another possible method for introducing new transportation systems is in the building of new towns, or in somewhat smaller developments such as new airports, or in amusement parks such as those run by Disney. Some short-range new systems have been successfully introduced, such as the shuttle in the Dallas-Fort Worth airport. On the other hand, Denver airport management's disastrous experience with an automated baggage-handling system has undoubtedly scared many people from massive demonstrations of vaunted new technology.

My preference, the free-market approach, would involve the introduction of various means by which motorists would pay something close to their true costs of travel. The fuel tax should be much higher to pay for the costs of safeguarding our fuel supplies and the costs of research into smog and global warming. There needs to be a fee with a much finer discrimination to compensate and control some of the effects of congestion and pollution. There is a growing consensus around the world that road-use pricing is the answer. Motor vehicles could be equipped with road-use meters with pre-purchased road-use units. These units would be "zapped" by transponders on utility poles on all except minor roads (see Figure 7.3). The units deducted would vary with place and time: high at times of congestion, in downtown areas, and other places where traffic "calming" is desired; low at other places and times. Parking fees at on-street meters and elsewhere would follow the same pattern, and could be collected electronically by the same on-vehicle meters. All parking, including that at businesses, malls, movie theaters, restaurants, etc., would be required by law to charge market rates.

The results of these various charging measures would be enormously beneficial. Traffic jams would completely disappear. Cities and towns would have enormous revenues, enabling them to reduce real estate and other

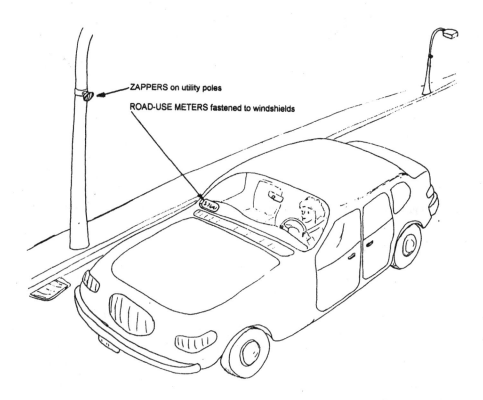

ZAPPERS on utility poles

ROAD-USE METERS fastened to windshields

**Fig. 7.3  Road Use Meters**

burdensome taxes.  Walking and bicycling would be fun again in suitable localities as they became popular alternatives for all or part of commuting and shopping.  Urban sprawl would be stopped because people would no longer have the incentive to drive "beyond the congestion".  People would use motor vehicles only for appropriate trips.  Buses would be able to travel faster, not being held up in traffic, and would be much more frequent.  Buses and rapid transit would no longer need subsidies, and could be taken over by private enterprise.

This would be a dramatic change.  At present, public transit authorities have difficulty contemplating any new technology more drastic than new lighting or fare collection systems.  If, however, such authorities became regulated private industries, encouraged simultaneously to serve the public and to make a profit, like telephone systems, they would have to consider all possible new technologies to stay competitive.  The next topic is a set of transportation systems that should, I believe, be developed.

## 7.4.   NEW TRANSPORTATION TECHNOLOGIES

The following three new forms of local ground transportation are choices from among many possibilities. They should complement each other. They should replace the automobile in many situations. This is not an anti-car message. The car is a wonderful, almost miraculous, device, giving people unparalleled freedom to visit people and places. The problem with our current transportation mix is principally that we frequently use the car in highly inappropriate ways.

### 7.4.1   Transportation Systems Based on Human-Powered Vehicles

The term "human-powered vehicles", or HPVs, applies to familiar bicycles and to various less familiar recumbent bicycles, tricycles, and four-wheeled vehicles. It also applies to human-powered aircraft, submarines, and so forth. The 200-meter speed record for an enclosed recumbent bicycle is at present 68.4mph (see Figure 7.4). A team of four riders in a similar machine crossed the U.S., coast-to-coast, in five days. It is not out of the question, therefore, that HPVs propelled by ordinary fit people should be able to travel at the 6-15mph that is the average speed of a large proportion of urban traffic. HPVs could thus be as significant a transportation mode in much of the U.S as is bicycling in the Netherlands and in China.

There have in the past been adventurous schemes using completely separate "guideways" with HPVs traveling on rails or other tracks.[3] Proposals for such systems are still being made. My preference is that of the majority of the bicycling public: for a marked half-lane at the right of regular roads, continuing through intersections. The best example I know of for such bicycle lanes and intersections is Christchurch, New Zealand.

Such provisions hardly constitute a "system"; they are necessary pre-conditions for HPVs to be used more widely, but are marginally sufficient only in flat countries. In cities like Pittsburgh, San Francisco, and Seattle, long steep hills are additional severe barriers to widespread HPV use. One solution to this problem is to designate some uphill streets as bicycle routes; to ban parking on the uphill side; and to install moving handrails as used on escalators along the side of the road. Bicyclists would grasp the handrail and be hauled up the hill as if on a tow rope (see Figure 7.5). There may be some need to devise some additional safety features. But even as baldly stated, the scheme would bring some ease and fun to the use of bicycles in urban areas.

**Fig. 7.4 The Cheetah team on September 23, 1992 in the Great Sand Dunes National Park, the day after breaking the land-speed record. From l-r: Chris Huber, professional cyclist; John Garbarino, Kevin Frantz, and James Osborn, formerly UC Berkeley mechanical engineering undergraduates.**

**Fig. 7.5 Uphill Escalator-handrail Assist**

### 7.4.2  Nonstop Buses

During winter cold spells, riders are often troubled (as they pedal along, glowing warmly) that people must wait at open bus stops, sometimes for fifteen minutes or more, for buses that cannot, because of traffic congestion, keep to schedules. The scheme I propose would be much enhanced by the elimination of congestion by the means advocated herein. It could be appropriate in regions where the number of potential passengers is not high enough to justify a full rapid transit system.

In the proposed scheme, each bus stop would be equipped with at least one four-person sit-down capsule connected to a heating or cooling unit. Thus passengers would no longer have to wait in the open in inclement weather. There would be room for a wheelchair or baby carriage. There would also be emergency buttons so that anyone feeling uneasy at being closeted with a stranger could summon a rapid-reaction force. Each capsule would be waiting at the start of a short track. Each bus would have a docking station for one capsule at the rear of the bus. At the approach of an appropriate bus, which would slow but not stop, the bus would allow a deceleration track to accept its capsule with disembarking passengers, and almost simultaneously the waiting capsule would be accelerated and transferred to the bus (see Figure 7.6). Each of these acceleration and deceleration tracks would be about ten yards long.

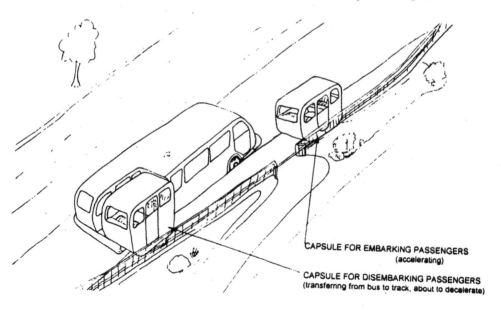

CAPSULE FOR EMBARKING PASSENGERS
(accelerating)

CAPSULE FOR DISEMBARKING PASSENGERS
(transferring from bus to track, about to decelerate)

**Fig. 7.6  Loading/Unloading for Nonstop Bus**

Would-be passengers would not have to wait in unpleasant conditions; passengers in transit on the bus would not have to endure the long stops every half-minute for passengers to embark and disembark, and average speeds would at least double and possibly triple. The system would be enormously less expensive than a rapid transit system and, with the greater utilization of the bus and driver, should be more cost-effective than the traditional system. It is highly desirable that only in very rare circumstances should passengers be left behind at the bus stop or on the bus. Therefore, the optimum size of the capsule needs to be chosen judiciously. Perhaps it should take six or eight passengers.

### 7.4.3  Universal and Personal Rapid Transit

Rail-borne, rapid transit systems have developed along a logical but counterproductive path following decisions made in the last century to use steel (or iron) wheels on steel rails. This technology gives very low rolling resistance but poor traction and braking capabilities. For safety, therefore, long distances must be maintained between trains. (The minimum "headway" is 90 seconds, equivalent to a two-mile separation for high-speed systems. The more usual minimum headway is five minutes.)

To carry many people, long trains of large vehicles must be used. Track, bridges, tunnels, and stations must be made large enough to handle these long heavy trains. A great deal of energy is required to accelerate trains leaving stations, and similar amounts are dissipated when stopping, even if only one or two people are embarking or disembarking. Thus, steel rail rapid transit systems necessarily involve, for the passengers, waiting what can be a considerable time for trains, and then coping with a large number of people trying to get off and on; and, for the municipality or management, extremely expensive construction and operation and large energy use per passenger-mile (higher than for automobiles).

Another technology was adopted by motor vehicles: rubber tires, giving generally very good braking capability. Early highways were still somewhat like railroads, with "on-line" stations: when one vehicle stopped, other vehicles in the lane would also have to stop. Traffic engineers installed multiple lanes, easing the problem considerably for motor vehicles (and to some extent for railroads), but multiple lanes did not cure the problem. Then the autobahn or expressway principle was invented: stopping on the highway was disallowed. To stop, vehicles had to leave the highway. A major increase in capacity of roads resulted.

This technology could not be transferred to railroads for at least two reasons: the braking distance is too long to allow at-speed switching and subsequent deceleration; and track switches cannot be operated fast enough and reliably enough to allow one vehicle in a string to be diverted to an exit ramp. In-vehicle switches were therefore designed: one vehicle can go straight ahead at a turn-off point at which the next vehicle, following closely behind, can switch to the exit or turning track. The usual method of doing this is to have vehicles guided by a rail or wall or slot on one side of the track. Vehicles scheduled to take a left or ahead direction at a fork would choose the wall (for example) on the left and vice versa. This choice could be made anywhere along a considerable distance upstream from the fork.

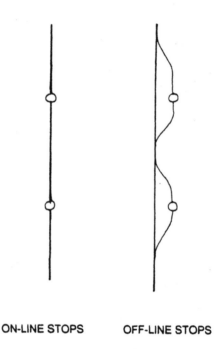

ON-LINE STOPS          OFF-LINE STOPS

**Fig. 7.7 Systems with on-line and off-line Stops**

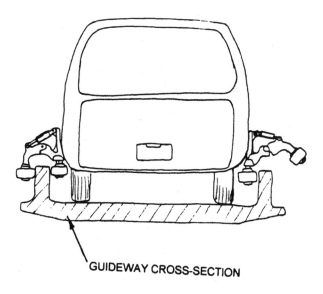

GUIDEWAY CROSS-SECTION

**Fig. 7.8  In-vehicle Switching**

Many investors and developers have applied these principles to new forms of urban transportation systems.  They lead to design and operating incentives completely different from those applying to railroads with on-line stations.  Because only short distances are needed to separate vehicles, there is an enormous increase in the number of vehicles or trains that can be carried per hour.  There is an incentive, therefore, not to form long trains of large vehicles but to use light, small individually powered vehicles carrying four to twelve passengers or the equivalent volume of freight.

This type of system has large benefits for passengers and for municipalities/ managers.  A vehicle taking on three or four passengers at one station, for example, Logan Airport in Boston,all of whom want to go to Lexington (a Boston suburb), can get routed directly there without stopping at a single station along the way.  This trip will be far faster than taking a car or going by car.  Construction costs for a narrow guideway taking light vehicles will be far less than for a rapid transit system.  The generic name given to this type of system is Personal Rapid Transit, or PRT.

### 7.4.3.1 Propulsion and Control

A public rapid transit system using many small vehicles would not be economically viable if each vehicle had to have a driver. All the systems that use the PRT principles have specified driverless vehicles under automatic control. From the engineering point of view, this becomes a very interesting challenge. Most developers have proposed using variable-speed motors driving rubber-tired wheels, with the speed responsive to the distance to the vehicle ahead.

When the author's group at MIT considered the propulsion-control problem over 25 years ago, we were concerned over the possibility of traction loss due to oil or water or ice on the track, and of wave action developing in a string of spacing-controlled vehicles. We were intrigued by the possibility of linear synchronous motors, but at the time they were not sufficiently developed. I had just walked up Mt. Washington (in New Hampshire) by a route that went under the cog railroad, and was impressed that this ancient system worked so safely in such rugged terrain in the worst weather in the world. We were granted patents on a modernization of the cog railroad: the cogs were driven by (rotary) synchronous electric motors, thus keeping one vehicle an exact number of rack teeth behind the vehicle in front. We moved the cog so that its axis would be vertical, engaging a rack on a low wall to the left of the track. The wall carries the electric-power rails and a "capture" surface to give positive engagement of the cog and positive guidance for the vehicle.

Another method of propulsion and control that avoids the uncertainty of wheel-track adhesion is the linear induction motor, used by the world PRT leader, Taxi 2000 (see Fig. 7.9). This innovative system has been developed by Dr. J. Edward Anderson of Boston University and the University of Minnesota.[4] It is scheduled to connect the Chicago Transit Authority's Blue Line with various locations in the Village of Rosemont in 1999. The linear-motor system is designed to give five-second headways between vehicles, sixty times better than most rapid transit systems. A license for this development has been taken up by Raytheon. It has been re-named PRT 2000. A test track began operating in 1995 in Marlborough, Massachusetts.

## 7.4.4   Single- or Dual-Mode Operation?

TAXI 2000 has captive four-passenger vehicles giving it the characteristics of a fast, highly convenient, short-wait, rapid transit system. Many developers have tried to incorporate dual-mode operation, whereby all the advantages of PRT would be conferred on vehicles that could also travel on the streets and highways.

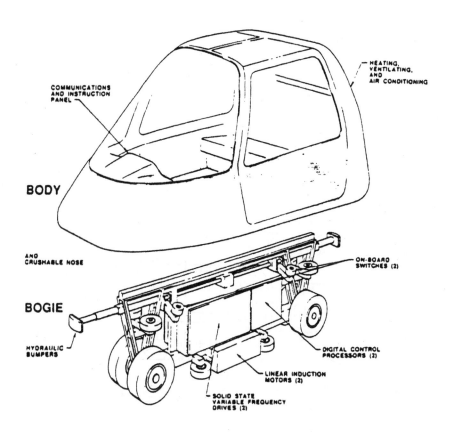

**Fig. 7.9 TAXI 2000 Propulsion and Control**

The danger of having privately owned and possibly poorly maintained vehicles on a close-headway guideway drove my PRT group to design flat cars or pallets on which street vehicles could be automatically loaded and removed. We called this the Palleted Automated Transportation (PAT) system[5] (see Fig. 7.10).

Such a choice produces considerable benefits together with additional costs. The benefits are that small buses could pick up passengers in a residential area, drive a short distance to a PRT station, travel downtown or to a shopping area, and drive off the system, distributing passengers at or very close to their destinations, and vice versa for the return journeys. The same loading/unloading mechanisms could also be used to move freight containers around the system. Low rates would apply at night, increasing the system's utilization. Many trucks would thereby be taken off the roads.

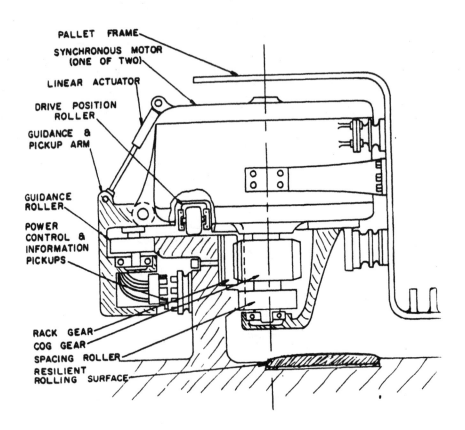

**Fig. 7.10 The PAT Propulsion-Control System**

The additional costs of a palleted system are connected with the cost of building and moving around the added material in the pallets, including the hold-down mechanisms securing vehicles and freight containers to the pallets. A pallet carrying a four-passenger mini-bus will obviously be somewhat heavier and larger than the bus designed to use the guideway on its own wheels, so that tunnels, for instance, might have to be slightly larger in diameter.

VEHICLES BYPASS INTERVENING STATIONS

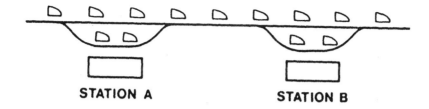

STATION A                              STATION B

**Fig. 7.11  TAXI 2000 Station (from a recent brochure)**

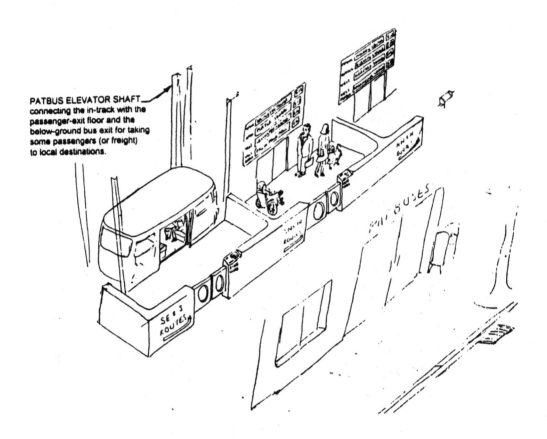

PATBUS ELEVATOR SHAFT
connecting the in-track with the
passenger-exit floor and the
below-ground bus exit for taking
some passengers (or freight)
to local destinations.

**Fig. 7.12  A downtown PATBUS Station**

### 7.4.5   The Future for PRT

The $40 million Chicago-Rosemont project has revived interest in PRT
worldwide. This interest is mainly, at present, in R&D. There is a one-million
dollar project at Bristol University in England, and other projects in Europe,
Korea, and Australia.[6] There seems little doubt that if the Chicago trial of PRT
is a success, no more traditional rapid transit systems will be built, and there will
be rapid development of increasingly sophisticated automated transportation
systems.

## 7.5 CONCLUSION

The U.S. has a reputation for allowing some problems to become an outrage before tackling them decisively. Traffic congestion and the urban sprawl and blight that it engenders seems likely to follow the same pattern. Courageous experiments can bring about courage in others to solve their problems. If the PRT 2000 development in Chicago is successful, governments around the world will be likely to introduce similar PRT systems. On the other hand, if any one of the many authorities considering introducing electronic road-use charging carries out a successful experiment of that system, and if it is seen to eliminate congestion, similar road-use-charging systems will spread worldwide with great speed. If they did, they would encourage many innovative transportation developments, including those reviewed here. Let the experiments begin!

### Notes

1.   Davidson, F.P. Figure given in private communication, 1995.

2.   Wilson, D.G., "A Subsidized Energy Binge," *The Washington Post*, 6/27/93, p. C7.

3.   Wilson, D.G., "Transportation Systems Based on HPVs", Fourth Scientific Symposium, International, 1992.

4.   Anderson, J.E., "Optimization of transit-system characteristics," *Journal of Advanced Transportation*, vol. 18, no. 1, 1984.
     Anderson, J.E., "The TAXI 2000 personal rapid transit system," *Journal of Advanced Transportation*, vol. 22, 1988.

5.   Wilson, D.G., "Palleted automated transportation - Developments at MIT," IATSS Research, *Journal of the International Association of Traffic and Safety Sciences*, Tokyo, 1989.

6.   Anderson, J.E., personal communication, 1995.

# 8

# Prefabricated and Relocatable Artificial Island Technology

**Ernst G. Frankel**
*Professor of Ocean Engineering and of Project Management*
*Massachusetts Institute of Technology; Senior Advisor*
*U.N. International Maritime Organization*

## 8.1. INTRODUCTION

Artificial islands have a long history, and are usually constructed by reclamation, which implies building up the island by depositing soil, sand, or other construction materials until the water surface is penetrated and an island surface created. To protect such reclaimed artificial islands, protection of the sides of the island by armoring with stone or concrete, containment of reclamation material within concrete walls or armored rubble mounds is usually employed. Similarly, to improve the footing of the island, foundation improvements by cement hardening of bottom material, sand and structural piling, or sand and gravel foundation carpet placements are used.

Such artificial islands have been used as storm barriers, offshore fishing bases, foundations for offshore gun placements for coastal protection, lighthouse foundations, and more since time immemorial. In more recent years, reclaimed artificial islands have been constructed to serve as solid waste depositories or fills, sites for toxic industrial activities, nuclear power plants, refineries, marinas, and airports, such as the mammoth landfill for Kansai Airport in Osaka Bay.

Reclaimed artificial islands take a long time to construct, often suffer from settlement, and become very costly if constructed in deep and/or hostile (exposed) waters. As a result, their use is usually restricted to water depths of 10-20 meters and for conditions where the island is largely sheltered and founded on a reasonably solid bottom foundation.

The use of reclaimed artificial islands has mushroomed in the last 20 years as industries and ports located on urban waterfront land were forced to move onto such islands offshore to:

1. remove desirable activities from the inner city waterfront;
2. provide access for larger ships;
3. free inner city waterfront land for recreational, residential, and commercial use (and taxable use);
4. develop facilities for treatment and disposition of waste; and
5. improve urban logistics.

Most major ports in Japan, Singapore, Korea, and other countries have moved in this direction. In total, over 67 artificial islands with a reclaimed land area of over 10,000 acres have been developed in Japan alone since 1972. There are similar opportunities in the U.S., such as in Boston Harbor.

Development of offshore terminals, particularly to replace the inner harbor (Chelsea Creek, etc.) oil, gas, and bulk terminals, which constitute major pollution sources, environmental, and safety hazards would free up hundreds of acres and feet of prime waterfront land for recreation, residential, and commercial development. This would eliminate the need for dredging the port, improve the tax base of the city, and permit road and pipeline arteries to be developed along the waterfront and/or in shallow water using low-level causeways or prefabricated sunken tunnels. In recent years, one or two tanker or barge mishaps per month have occurred in U.S. ports. It would be particularly unfortunate if such a disaster happened soon after the Boston Harbor cleanup.

There are approximately 580 acres of prime waterfront land and about 1.8 miles of Boston waterfront, in the Chelsea Creek, Mystic River, which could be released within a few years if offshore terminal facilities were provided at locations such as the Brewster Islands with adjacent deep water access and shallow banks for reclamation.

Tank farms would probably be left in place in the Chelsea Creek initially and the new deeper draft offshore terminals connected to existing tank farms by pipeline (except LNG). Similarly, the Mystic Container Terminal facilities could

be moved. Once this ship traffic has been eliminated from the inner harbor, inner harbor waters could be used to construct:
1. a low-level causeway instead of the Third Harbor Tunnel, at a saving of more than 50% of the harbor tunnel cost and with more ready expansion and connection potential;
2. a causeway or partially sunken Central Artery bypass which connects the Southeast Expressway with the harbor tunnels, Route 1, and Route 93 north, and also provide two or three access ramps from the city. All of this could be constructed without interfering with inner city (or cross-city) traffic. This would separate the south-north expressway traffic from city traffic.

With all the connections made, the elevated part of the Central Artery could be removed and the land under it used for recreational and commercial development.

The cost of a causeway-type road above water instead of a depressed Central Artery with connections and unlimited expansion (widening) potential would cost less than half as much as the depressed Artery and serve a much wider purpose or function. Furthermore, such an approach could be implemented in about half the most optimistic time projected for the completion of the depressed Central Artery and the Third Harbor Tunnel and could be expanded at any time by widening or branching.

Artificial offshore island port terminals are used efficiently in many parts of the world. They are used as central deep draft coal ports and central stockpiles to serve many coal-burning power plants located in inner cities that, as a result, were able to eliminate the need for coal stockpiles near the power plant, reduce the total coal inventory, as well as coal transport costs and environmental impact.

The central artificial coal island terminal is expected to pay for itself from inventory and operational cost savings within less than three years and, if the value of the inner city coal stockpile area is considered, will actually result in a huge financial windfall plus an increased tax base for the cities in which these power plants are located. Similarly, a central offshore deep draft artificial island oil terminal would generate large oil transport and inventory cost savings, while releasing about 450 acres of prime waterfront land to other more value- and tax-generating activities. The real estate value of this land is a multiple of the cost of artificial islands and relocating tank farms and pipeline systems.

Refineries, petrochemical plants, hospitals, prisons, waste treatment centers, ports, airports, and many other activities are increasingly being placed on offshore islands for economic, environmental, logistic, and strategic or political reasons.

Artificial offshore islands, although expensive, are economically very attractive. Their costs usually vary from $50-250 per square foot (depending on water depths, geophysical, environmental, and other conditions), and they compare favorably with the value of waterfront land in most locations.

Since 1970, over 100 reclaimed artificial islands have been built worldwide with a total area in excess of 28,000 acres or about 41 square miles. They have contributed not only to the expansion of waterfront land, but more importantly, to environmental protection, access to deep water transportation, and effective urban development by facilitating separation of urban industrial and commercial, residential, and recreational activities.

## 8.2. PREFABRICATED ARTIFICIAL ISLANDS

Prefabricated artificial islands have been largely developed for use in offshore petroleum operations, although Texas tower platforms were used during World War II for coastal protection, and similar columned structural supported or floating platforms have been developed for a large variety of uses. Figure 8.1 shows a very large deepwater gravity column-supported oil production platform used in the North Sea, while Figure 8.2 shows a number of different types of prefabricated platforms. As shown, these vary from simple barge-type platform structures as used for a floating container terminal in Valdez, Alaska, to tension leg, gravity caisson, column-supported or jackup-type platforms. All of these platforms are designed to be prefabricated, floated into place, and assembled or erected at the site. They are made of reinforced concrete, steel, mixed or composite material construction. Very large floating storage facilities, such as those shown in Figure 8.3, are also of increasing interest (4 x 250,000 = 1,000,000 tons of petroleum or grain). A one-hectare (2 1/2 acre) floating island, named Aquapolis, was used during the International Ocean Exposition at Okinawa in 1975. Since then, artificial prefabricated island technology has greatly advanced and unit costs have declined.

## 8.3 PREFABRICATED, FLOATABLE ISLAND STRUCTURES

The concept is designed to be built in a shipyard, usually in modules of between 20m x 40m to 120m x 240m, and assembled afloat before or after towing to the operating site. The island is completely outfitted and may also carry superstructures consisting of hangars, buildings, etc.

The structure consists of an array of gravity caissons of unique design, equipped with bottom skirts which enclose voids and support a column-stabilized platform. The caissons are open at the bottom and are designed with a

Fig. 8.1  Large Concrete Gravity Column Supported Oil Production Platform.

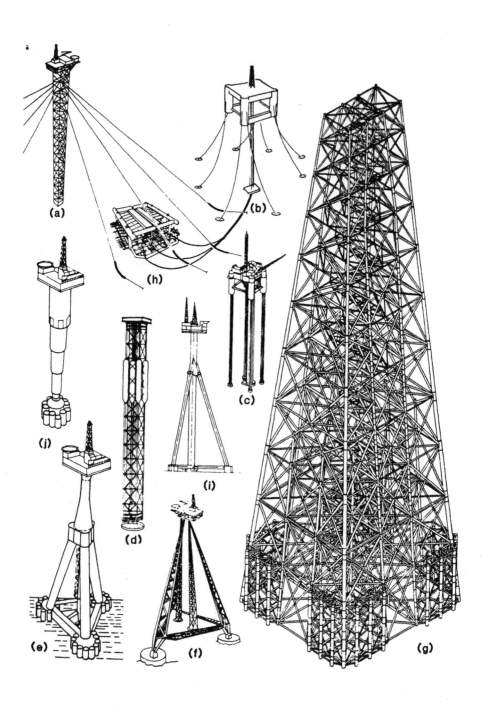

**Fig. 8.2 Prefabricated Platform Types** (source: McDermott International)

**Fig. 8.3 Floating Oil/Bulk Storage Facility**

circumferential skirt as well as vertical separation baffles in the open void. The skirts and baffles have lengths which depend on the bottom soil conditions and would usually be 14m in length. The skirts and baffles project from the reinforced bottom of the caisson. The caissons also have an inner bottom and a ballast or cargo space as shown in Figure 8.4.

The bottom caissons are supplied by a pressurizing system which injects sand or cement slurries into the voids under the caissons contained by the bottom protruding skirts which are normally driven into the bottom soils by gravity and then maintain the pressure to facilitate the leveling of the array of gravity caissons which carry the stabilizing columns. These in turn carry the platform box girders.

The bottom caisson and the columns they support would usually be arranged in a grid pattern. In turn, the columns support a box girder which supports a multi-deck platform structure as shown.

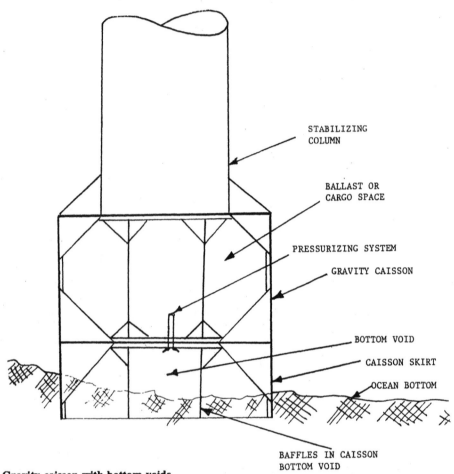

**Fig. 8.4  Gravity caisson with bottom voids.**

The deck structure with the superstructure is constructed on a floatable box girder. This would normally be floated on top of the sunk caisson array as shown in Figure 8.5, which indicates a caisson array made of large-diameter cylindrical caissons. These can be deballasted to mesh with, engage, and lift a box girder-supported superstructure as shown in Figure 8.6.

The combined structure, attached by welding, can be used as a floating, column-stabilized platform, and towed to a shallow water location to be placed on the ocean bottom as a gravity caisson array with a small water plant. In case water depths where the structure is to be operated as a gravity caisson array-supported, small water plant, stabilized structure are very deep, then a second gravity caisson array can be placed under it, meshed and attached to the system array attached to the box girder as shown in Figure 8.7.

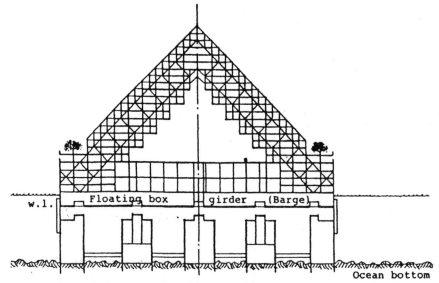

Box girder supported superstructure floated
on top of ballasted (sunk) support array

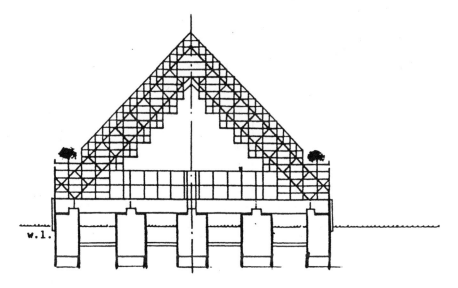

Deballasted support system array lifts box
girder supported superstructure which is
attached to support system array by welding

**Fig. 8.5** Typical superstructure on Box Girder Module placed on column stabilized support
system.

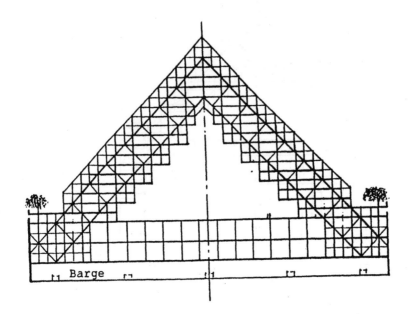

Section:

A-frame superstructure built over the standard
structure.

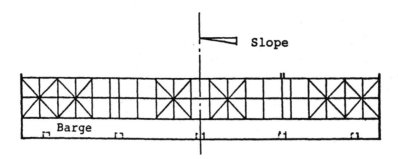

Section:

Standard structural module for the airport runway,
taxiway, aircraft parking, auto parking, etc.

Fig. 8.6 Box girder barge structure.

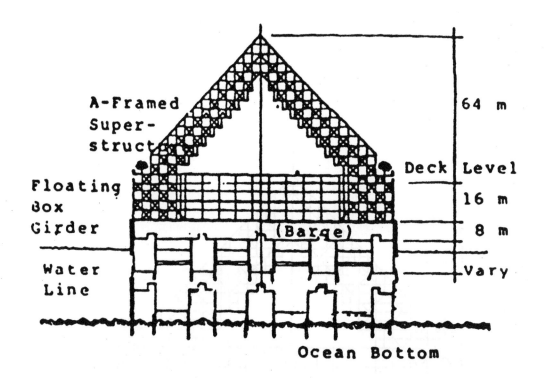

**Fig. 8.7** Deballasted support system array lifts box girder supported superstructure which is attached to support system array by welding.

The column caisson array support structure can be configured with separate bottom gravity caissons serving as submerged displacement vessels when the platform is supported by connecting columns and arranged as a semi-submerged multi-maran as shown in Figure 8.8.

The stabilized-column gravity caisson arrays-supported box girder and superstructure modules are connected by semi-flexible couplings which permit limited relative motions in two or three degrees of freedom to form airport, heliport, recreation, or commercial artificial islands. The modules shown present an A-frame modular superstructure which can be designed to serve as an office building, hangar, terminal, or industrial facility (Figures 8.9 and 8.10).

The modules can be connected afloat by adjusting the ballasting or after placement on the ocean bottom, by adjusting the pressure and volume in the leveling and elevation voids between the circumferential skirts under the bottom gravity caissons or column gravity caissons, so as to assure equal levels of the platforms for effective coupling. A special hydraulic/mechanical coupling device

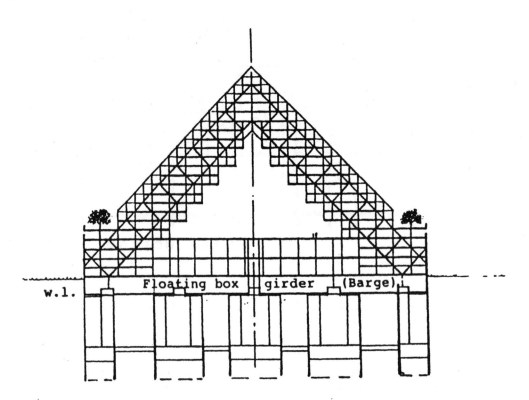

w.l.

Floating box   girder   (Barge)

Ocean bottom

**Fig. 8.8 Typical superstructure on box girder module placed on semi-submerged multi-maran support.**

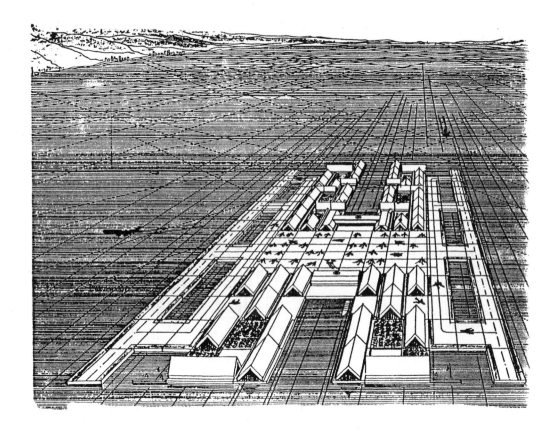

**Fig. 8.9  Aerial view of prefabricated, floatable island airport.**

1. 3,600m runways

2. Taxiways

3. Aircraft parking

4. Terminal building

5. Hangers

6. Container port
   Ferry terminal

7. Industrial plants

8. Worker's housing

9. Fresh water
   reservoir

10. Residential area
    shopping and
    recreation

11. Marina and
    restaurants

12. Staff apartments

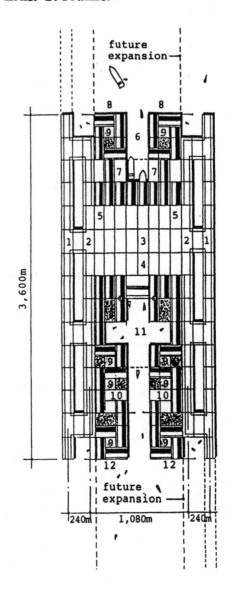

Site Plan:

**Fig. 8.10  Floatable, prefabricated island airport layout**

has been developed for this purpose which includes a damping and tensioning capability. The multiple decks under the runway are designed to accommodate passenger, baggage, food and other service supply systems, completely eliminating service and other operations on the runway and taxiing area, as shown in Figure 8.11.

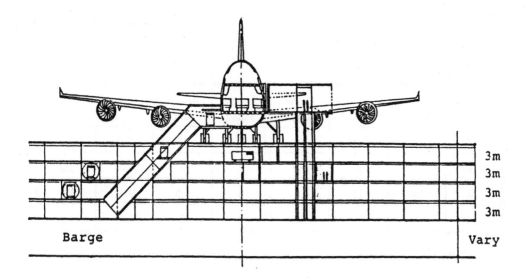

Section:

Four levels below the airfield level will be used for passenger and baggage transfer.
Hydraulic elevators will carry passengers to the lower levels.
45° conveyor system or a hydraulic system will carry baggage to the terminal building.

**Fig. 8.14  Underdeck service operations**

## 8.4 BEHAVIOR OF THE FLOATING, PREFABRICATED ISLAND MODULES

The motions of a floating column stabilized box barge girder (240m x 120m) at different drafts with gravity caissons the same diameter as columns and columns on a geometric grid were computed to have heave, surge, and pitch motions as listed for a typical module in Figure 8.15.

| | Wave Characteristics | | Predicted Motions | | | | |
|---|---|---|---|---|---|---|---|
| Draft | Height | Period | Heave Ampl. | Steady Value | Ampli-tude | Pitch Ampl. Radians | Degrees |
| 100 | 30 | 11 | 1.1 | 8 | 10 | 0.0 | 2.0 |
| 100 | 30 | 8 | 1.0 | 7 | 4 | 0.04 | 2.0 |
| 70 | 30 | 11 | .8 | 6 | 1.5 | 0.02 | 1.0 |
| 100 | 10 | 8 | .02 | 0 | 2 | 0.03 | 1.2 |

**Fig. 8.15 Predicted motions of the floating box barge girder 240m x 140m on caisson or column supports (15m diameter)**

Families of modules with different dimensions, grid spacings, and stabilizing column diameters were developed to permit selection of the best design for particular environmental conditions.

The towing resistance of column-stabilized modules was similarly determined for all configurations of island modules (Figure 8-16). Analysis and model tests have also been performed to determine the relative motions of the box girder barge and a submerged array of gravity caissons with mounted stabilizer columns and how the ascent of the latter can be controlled for effective mating of the two larger components. Similarly, the coupling of completed floating modules has been analyzed and effective coupling systems have been developed. A scale model (12m x 24m) is under investigation to test the results of the analysis.

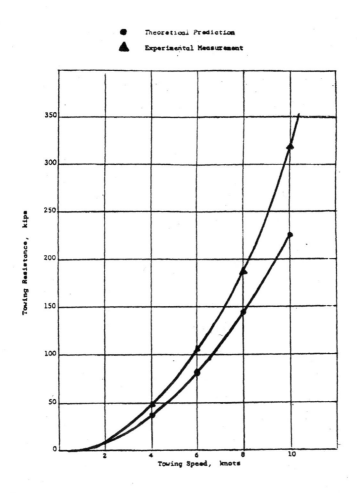

**Fig. 8.16  Towing resistance (120m x 240m / 15m columns)**


## 8.5.    STRUCTURAL DESIGN AND MODULE WEIGHT

Various module configurations have been designed and alternative support array systems have been developed with different grid spacing and stabilizing column diameters.    Figure 8.17 provides a summary estimate of steel and outfitting weights for modules designed for various purposes, including buildings and other facilities.    A basic (parking) module, on gravity caisson-supported stabilizing columns, for example, has a steel weight of 31,911 tons and with all outfitting (pressurizing, electric power generators, electrical, ballast, water, etc. systems) the weight increases to 37,810 tons.

The design in all cases permits both floating or gravity operation of modules and the longitudinal and transverse coupling of modules to widen or lengthen the island.

| Steel Weights | Basic Structure on Box Girder | On Support Array | | | |
|---|---|---|---|---|---|
| | | Deep Barge | Semisub | Gravity Caisson | Semisub Multimaran |
| Terminal Super. | 29,352 | 41,152 | 41,152 | 43,652 | 45,452 |
| Hangar | 26,630 | 38,430 | 41,430 | 40,930 | 43,730 |
| Parking | 17,611 | 29,411 | 32,411 | 31,911 | 34,711 |
| Aircraft Pkg. | 20,546 | 32,346 | 35,346 | 34,846 | 37,646 |
| Runway | 20,900 | 32,700 | 35,700 | 35,200 | 38,000 |
| Business-Hotel | 29,722 | 41,522 | 44,522 | 44,022 | 46,822 |
| Residential | 29,722 | 41,522 | 44,522 | 44,022 | 46,822 |
| Including Outfit Weights (outfit weights include intermediate decks where applicable) | | | | | |
| Terminal Super. | 42,980 | 56,480 | 49,980 | 57,980 | 60,880 |
| Hangar | 33,210 | 46,710 | 50,210 | 48,210 | 51,110 |
| Parking | 22,810 | 36,310 | 39,810 | 37,810 | 40,710 |
| Aircraft Pkg. | 23,480 | 36,980 | 40,480 | 38,480 | 41,380 |
| Runway | 24,100 | 37,600 | 41,100 | 39,100 | 42,000 |
| Business-Hotel | 43,280 | 56,780 | 60,280 | 58,280 | 61,180 |
| Residential | 42,300 | 55,800 | 59,600 | 57,600 | 60,500 |

240m x 120m section in tons (6m deck height). Assuming 2m x 6m decks under runway dock.

Fig. 8.17 Summary estimate of steel and outfitted weights (15m diameter columns - 30m grid)

## 8.6 ECONOMIC VIABILITY

The construction and erection costs of a basic module of 120m x 240m without superstructure, but otherwise fully outfitted, are estimated to be $38-55 million, or $1319-$1909 per square meter of platform (or $43-$572 per square meter of 6m high deck space - if intermediate decks are introduced for a 3m deck height, then these costs are halved), costs which compare favorably with those of reclaimed islands and, for that matter, with those of other prefabricated floating or floatable artificial island concepts.

Construction and erection times are estimated to be significantly lower than those of other types of artificial islands. Furthermore, the concept is expected to be able to sustain semi-exposed environmental conditions. It can be designed with rigid or semi-rigid skirts and can be placed on various types of bottom soils varying from soft silts and clays to sand/gravel and even rock, when semi-rigid skirts are used to contain leveling concrete and sealing injections. The skirts are slightly tapered to permit ready refloating of the caisson-supported structure or to permit mating of caissons to each other.

## 8.7.    CONCLUSION

The concept of a prefabricated, relocatable artificial island as described appears to offer opportunities for the economic use of offshore sites for numerous activities. At this time, the concept is under consideration for the construction of prefabricated, relocatable earthquake-proof hospitals and emerging power supplies in Japan, offshore airports in Japan and Israel, offshore hotels, recreational and cruise terminal centers in various locations in the Caribbean, as well as several potential industrial applications for waste incineration and treatment and petrochemical manufacture.

# 9

# The Command Tactical Information System: Military Software for Macro-Engineering Projects

**Thomas McInerney**
*Lt. Gen. (USAF, Retired); former Assistant Vice Chief of Staff, United States Air Force*

## 9.1 INTRODUCTION

A comparatively new technology is emerging to which the term "engineering" is applied: "system engineering". Most macro-engineering projects have a reasonably concrete definition at the outset and are expansive and expensive. Examples range from the Great Pyramid of Giza to the Channel Tunnel project. Disciplines supporting more traditional kinds of engineering, such as mathematics with its related mathematical models and physics with its related levels of certainty, have not supported system engineering to any appreciable degree. Large-scale system engineering projects, unsupported by disciplines like thermodynamics and strength of materials, fail at an appalling rate. Yet, this new engineering is shaping the world in which we live, sometimes in ways more subtle and profound than more traditional macro-engineering projects. New management and developmental approaches to system engineering may improve the chances for success.

These approaches were used in a highly successful macro-system-engineering project started at Elmendorf Air Force Base in Alaska to bring together for the commander and his staff relevant information about all the joint services. The new technology is known as the Command Tactical Information Systems (CTIS). Its reach was later extended via the North American Air Defense Command into Canada. Later still it was extended into the Office of the Secretary of Defense, and again into supporting joint simulated exercises; then into Korea to support the Joint Forces Air Commander. Finally, it reached the Supreme Allied Commander. In a parallel process, it was also adapted for use in the civilian sector to support rescue operations and disaster recovery.

By any measure, this project has had far-reaching effects and has shown great adaptability of application. In addition to its wide geographical extension beyond the military into civilian applications, the project has influenced the development of new software engineering technology, such as the migration to client-server and intelligent messaging technologies.

## 9.2 EXPERIENCES

Building CTIS was a unique experience. In part, it was an outgrowth of earlier initiatives in the Pacific in which I participated as a Commander in the Philippines, Okinawa, and seven operations exercises in Korea; then in Hawaii where it was used in Constant Watch, an operational intelligence system. Later, it was implemented in Korea in the Hardened Tactical Operations Center along with other very important command and control elements. CTIS was also used by the 3rd Air Force in the Libyan raid. During this same time, the world anxiously watched the Soviet Union during the "cold war" as massive Soviet forces were deployed across the East German border.

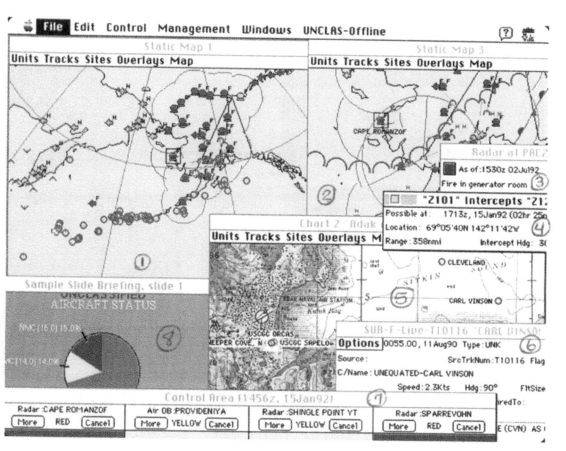

Fig. 9.1 A melange of executive information which can be distilled and displayed on the Commander's situation awareness screen. In various display windows from upper left: (1) A vector map of Eastern Siberia and the Alaskan NORAD Region depicting friendly radars (blue) and approximate radar coverage (green circles) (a failed radar in red with red circle of coverage); friendly (blue) and enemy (red) ships, submarines and aircraft and their approximate speed and direction; enemy air fields (grey); (2) A vector map of southwestern Alaska depicting failed (red) radars, one at Cape Romanzof, an AWACS aircraft (green circle of approximate radar coverage shown as well as direction and relative speed) filling the breech left by the Cape Romanzof failure; cloud coverage, light snowfall and a stationary front also depicted; (3) Indication of a failure of PACZ (Cape Romanzof) radar due to fire in generator room; (4) An intercept pairing including range and heading; (5) A bit-mapped map of the vicinity of Adak Naval Air Station overlaid with the location of Coast Guard and Naval vessels (e.g., the carrier Carl Vinson); (6) Speed, heading and other detailed information on the Carl Vinson; (7) a window for displaying the most pressing problem or warning advisories for the Commander (two red warnings, two yellow cautions); (8) A pie-chart briefing re Aircraft Status (red indicating NMC, not mission capable).

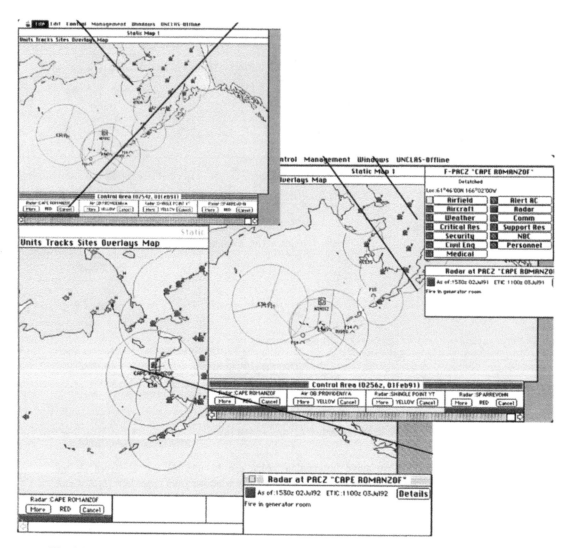

Fig. 9.2(a)    An unusual or emergency event occurs.  In this case, the radar at Cape Romanzov
                fails.  The radar icon turns red and at the bottom of the screen, the red alert appears.

Fig. 9.2(b)    Successively greater detail on the status of the site is obtained, if desired.  First 13
                general categories of the site's status may be displayed.  Second, a short description
                of the specific probelm can be displayed.  Even more information can be obtained from
                linked reports and still more detail can be obtained by interactive e-mail dialogue with
                personnel on scene.

Fig. 9.2(c)    The response to the failure is expedited; the AWACS fills the gap.

## 9.3    OVERVIEW OF THE DEVELOPMENT PROCESS AND LESSONS LEARNED

Operations in Alaska where I had command provided an opportune time to develop a system that could focus and unify the disparate commands. Programmers and developers, together with operators (i.e., users), purchased Commercial-Off-The-Shelf (COTS) software and hardware to build a system from technology available in the marketplace. These three elements (developers, users, COTS offerings) later became the accepted way to develop large-scale systems. This new development technique required a very evolutionary process.

The process worked like this: We would have an exercise which stressed the CTIS system, for example, a major Joint Chiefs of Staff exercise. We would use and test the software, the hardware, and the concept of using information technology in the fighting arena, and learn from the exercise. Then we immediately made changes to the software or the processes, adding additional equipment and functionality that industrialists were bringing out, and bringing the operational viewpoint into the changes.

The changes were made within weeks and those using the system, the operators, became directly involved with the developers and software programmers. This was a unique way of doing things at the time, but it was imperative that it be done this way. Programmers were allowed great flexibility of access to users so that they could begin to understand what the war fighter was talking about and how he was thinking. Good ideas came from people in other commands, and the best ideas were incorporated into the system.

## 9.4 WHAT WE LEARNED

Here are some of the lessons learned:
1) **Any system implemented must work.** Governments and the private sector have bought systems that do not work, so the system must work for the operator, the user, the customer.
2) **It must be user-friendly.** Generals must be able to understand and use the system, not merely technicians, computer scientists, and programmers. Young airmen who have no computer experience must also be able to use it. The Macintosh computer proved to be very helpful in this regard.
3) **It must be fast and responsive.** Response must be obtained from the system within two seconds or less.
4) **It must be low cost**, which is why commercial products are so important. Developing proprietary systems are not affordable. Go out, look at the

marketplace, bring it in and adapt to it, and then refresh the system as new technology becomes available. New chips are now being developed in less than 18 months, sometimes less.   This is completely revolutionizing the way things are done. It is driving costs down and we must adapt development processes and acquisition techniques to accommodate these changes.

5) **It must be inter-operable.**  In other words, the system must have an open architecture.  Army, Navy, Marines, Air Force, Coast Guard, commercial, civil sector, states, federal, must all be able to operate together.

6) **It must be secure.**  Nothing can be completely and perfectly secure today.  Systems must have security and there must be different safeguards from a System High to a System Low.

From these six characteristics, it evolved that future systems to be developed must:

1) **Have an open systems architecture,** and
2) **Use Commercial-Off-The-Shelf software and hardware.**

We must learn from systems such as CTIS.  We can adapt CTIS to controlling disasters, as in the Alaskan oil spill.  We ought to be able to use such systems to help during earthquakes, typhoons, hurricanes.  We can adapt it to the marketplace of running cities, for example, in London, New York, Los Angeles. Such a system can be used to see where ambulances, the police, and fire apparatus are currently located.  Everything required to run a city can also utilize CTIS-like technology.   Thus, CTIS was perfectly suited for tackling other problems.

## 9.5 ARRIVING IN ALASKA:  THE PROBLEM

What I found when I arrived in Alaska was a DOS computer operating system used at Wing level for Air Defense scheduling.  It was essentially the only computerized support for operations, and it did not provide a total picture of what was going on in the theater.

Facing us across the Far East Military District, particularly down at Kamchatka, were long-range Soviet aviation forces capable of swinging north toward us and penetrating our air defenses on a monthly basis.  It was necessary to launch airplanes at 2:00 A.M. out of places like King Salmon and Galena, on short runways, in temperatures of -40°F to -50°F, fly single-seat fighters carrying a maximum load, and then refuel.  AWACS were needed for tactical support.

And all of these components needed to be brought together. A picture of everything had to be available for all the forces in the alert sites.

## 9.6 EVOLVING A SOLUTION

A system was needed that would show a "red over blue" view, meaning a view of enemy forces and friendly forces. This system was called "3DS" (Digital Display System) which became the keystone of CTIS. 3DS provided the NORAD Region Commander with early warning of approaching Soviet airplanes, and field units could also see these forces well in advance.

This also gave maintenance personnel time to specially prepare the runways to ensure that they were in perfect condition so we were always ready and able to launch within 15 minutes. The situation was different when there was a blizzard and everyone knew that airplanes would not be coming. However, when airplanes were approaching, and a launch was required because of advance warnings given by 3DS, then things could be done more safely. The safety of air crews was paramount and 3DS enabled this safety.

Soon we also linked into the Army division so that both commanders had the ground picture of any deployment and the status of all U.S. forces in Alaska. Then it was linked into JOTS, the Navy's Joint Operational Tactical System, which provided a picture of all naval forces, surface and subsurface, red over blue. This was important because division command was interested in where submarines were located, as they served as "tippers." Submarines from Shimya or other sensitive areas gave us days of warning as to pending actions, so we could position ground forces for security. In the midst of the cold war, such "tippers" were needed to give a picture to the fighters and people who needed to act. We also tied in the all-source intelligence information to 3DS so commanders could see the red forces superimposed over the blue forces and make coordinated operational decisions. 3DS was near real-time, distributed over the command, and visual. This gave a verifiable picture which was vital then and for future use.

## 9.7 OVERSIGHT OF THE EXXON VALDEZ OIL SPILL

Another example of the power and flexibility of 3DS and CTIS was its use during the Alaskan oil spill caused by the *Exxon Valdez* which ran aground on Blye Reef. Using CTIS, we obtained a picture of what was happening. The recovery effort involved more than ten thousand workers in Prince William Sound. There were more than twelve hundred ships in the effort, some very small. For safety and management of the crisis, we had to know where every one of them was positioned. This information was supplied to my command headquarters. In turn, we supplied it to the Coast Guard Commandant's office in the Pentagon. The CTIS technology was utilized in crisis to help eliminate gridlock, to improve the efficiency of the rescue efforts, and to provide needed vision to the crisis overseers.

## 9.9 SUMMARY

CTIS provided a number of lessons:
1) It demonstrated a way for the Department of Defense to improve its acquisition system, information technology, and theater battle management.
2) It revealed improved developmental processes in which the programmers and operators (the customer in the commercial world) are involved in the overall developmental and change process.
3) It showed the power, speed, and effectiveness of using COTS technology, software, and hardware.
4) It predicted the move from mainframe to client server architecture, as Macintoshes were distributed throughout the command and functioned as clients to VAXs being used as servers.
5) It showed that technology can be continually upgraded.
6) It showed that we could integrate in a command and control system both joint forces and civilian organizations (Army, Navy, Air Force, Intelligence, Coast Guard, FAA, civilian).
7) It showed that such a system can be, and must be, user-friendly.
8) Finally, it showed that all of this can be done in a cost-effective manner.

In four years, we spent only $10,000,000 on the whole system for development, deployment, and integration. That is an incredibly low amount of money for such a scale of effort. Other monies were used in CTIS for other programs, but as far as deploying and developing CTIS, it required $10,000,000 in four years. It demonstrated that if the principles just outlined are used, it can be done -- it was done. It was secure; it was interoperable; it met all the criteria that had been laid out.

## 9.10    EPILOGUE

### 9.10.1  Korea

Where did CTIS take us?  Well, it took us into Korea, where we were able to deploy quickly, within four months, a rapid redesign of the CTIS system which enabled Lieutenant General Ron Fogelman, then 7th Air Force Commander (later promoted to General and then Chief of Staff of the Air Force) to have a usable system for aiding the execution of an air war in Korea in a joint (Army, Navy, Air Force) and combined (U.S., Korea, United Nations) responsive setting.

Meetings in Korea among General RisCassi (Supreme Allied Commander), Lt. Gen. Fogelman, Lt. Gen. Fornell (Commander of Electronic Systems Center (ESC)), and Tom O'Mahoney (Director of Command and Control Systems at ESC) resulted in the go-ahead for an ESC team to quickly place a prototype system tailored to the needs of Korea.  Earlier multimillion dollar attempts had failed or stagnated.  Thus, history indicated the effort would be high-risk.  The earlier failures placed severe restrictions on the acceptable time frame for success.

Requirements, different for Deep Battle Synchronization in Korea than for CTIS in Alaska, were articulated and defined.  A concept of operations was sketched out; a technical design was formulated; a management plan was developed; an implementation plan was developed; a plan for migration to open systems was completed.  The prototype system was deployed, integrated, and running in Korea in four months.  The adaptability of the CTIS system was confirmed.  Its cost-effectiveness was confirmed.  The process worked.

Once in use, the system provided a significant side benefit.  Other units in the military could monitor the situation in Korea so that reserve augmentees could get and maintain an overall vision of the conflict prior to deployment, thus effectively eliminating the time required for commanders to come up to speed once on scene.  This same kind of vision improved the prioritization of critical logistics support.

### 9.10.2  National Performance Review

Next, I moved to the National Performance Review and the Defense Performance Review initiatives of the Clinton-Gore administration to implement their pre-election pledge to reinvent government.  We used many of the management and development principles evolved from the CTIS experience in the Defense Performance Review, the highlight of which is the medical initiative known as the Medical Defense Performance Review (MDPR), headed by Brig. Gen. Peter

Hoffman, M.D. and ably assisted by John Evans and his MDPR team, ESC, and MITRE.

They have extended the CTIS management and development principles into the electronic patient record and telemedicine.

The Provider Work Station (PWS), originally started in Alaska and now being worked on at Scott Air Force Base, is being teamed with telemedicine. The PWS gives the health provider a graphically oriented, easy-to-use, computerized patient record to facilitate the patient encounter. A very powerful side benefit from the patient encounter is that the data captured is available for us in patient-level cost accounting. Thus, for the first time in the Department of Defense, cost decisions in medicine can be driven by patient-level cost accounting. In this way, medical treatment options can be correlated to medical outcomes to select treatments by cost-benefit. In the former absence of this kind of data, the obvious option for controlling the cost of medicine was to reduce the use of costly procedures and medications, because in many cases we did not have information available to support a decision based on effectiveness as a second factor.

Telemedicine technology was initially used to support the executive management in Air Force medicine. In our prototypes, we have used much lower bandwidth communications than traditionally, and thus have achieved a much lower cost. This has enabled us to use the technology much more vigorously, effectively, and globally. Lt. Gen. Andrew Anderson, Surgeon General of the Air Force, conducts a global executive medical management meeting weekly, which has greatly increased his ability to effect responsive decisions by foreshortening the decision time horizon.

As we combine the PWS with telemedicine, we used the same basic principles as used in CTIS. It must work; it must be user friendly; it must be fast, responsive; it must be low cost; it must be interoperable; it must be secure. These principles will lead us not only to play a major role in Department of Defense medicine, but in the entire U.S. health care system.

Our Medical Defense Performance Review efforts provide a major opportunity to reduce the cost and increase the effectiveness and reach of medical care. The current budget review and gridlock over Medicare and Medicaid is heavily based on the concern that medical costs will go to twenty percent of the gross national product as we enter the twenty-first century. But, our efforts can show the way to reduce the cost of medicine to nine or ten percent of the gross national product, figures that are more in line with those of other developed nations. So, I commend those working on these MDPR efforts, based on the principles developed in the CTIS experience, and say: "Keep up the good work!"

# 10

# Prospects for the Next Century:
# Survey and Suggestions

**Thierry Gaudin**
*Ingénieur-Général des Mines; Président, Fondation 2100, France*

Forecasting the next century requires a global view; no place in the world is completely independent of others. No profession is self-sufficient; all of us have to look at the planet as a whole - and on an interdisciplinary basis. As we contemplate the third millennium, I believe it is necessary to look back at what happened in the second millennium. Of course, the evolution of technology is quite unpredictable, but my thesis, and I say it just now at the beginning, is that if you look globally at the technical system in which a civilization works, you can draw a general "schema" of technical-system changes.

The late and lamented Professor West Churchman admonished us that to constitute a "system", man-and-machine contrivances had to have a discernible **purpose.** The responsible selection of goals implies at least a rudimentary knowledge of history and a sense of the requirements which the future will impose. It is, therefore, the intent of this chapter to outline a perspective in which policy and program options can be viewed synoptically; there is at least a

wide consensus on the social and environmental ills which we face.   If
"sustainable development" is to be achieved on a sufficient scale and within an
acceptable timeframe, commentators and analysts will have to approach the
delicate and difficult terrain of **implementation**.

In the "tightly coupled" world prefigured by Jay Forrester, it has become
essential to focus on the improved deployment of the power of large-scale
engineering.   Useful slogans such as "sustainable development" must now find
their practical place in the "real world", lest festering problems such as youth
unemployment, deforestation, and soil depletion remain without operationally
feasible solutions.

We are confronting the third millennium in a state of world-wide
intermingling.  During the last fifty years air transport has made long distance
travel and emigration a possibility for hundreds of millions of people.  Previously
people lived in their own particular territory, the source of nourishment in ancient
agrarian society, which we still jealously cherish as a memory. For the first time
since the invention of sedentary agriculture, Chinese, Indians, Africans, Mexicans
all find themselves nomads and part of a Diaspora spreading their chromosomes,
their music and their cuisine throughout the world. From now on it is not a
question of whether this or that civilization will be dominant in this particular
region but of understanding how our variegated and unique world civilization will
evolve.  Already we all have the same cars, the same washing machines, the same
telephones. However, we can learn much from the diversity of past experiences.
By taking a look at the world of techniques and their relation to the civilizations
in which they emerged, we can weigh up the chances of this world civilization,
consider the mistakes not to be made and define the conditions for its success.


## 10.1  WHAT'S NEW IN THE LAST THOUSAND YEARS?

If we travel back in time 1000 years to the 10th century, Europe was not the most
advanced civilization.   Islam had developed teaching, perfected medicine,
invented algebra, translated and disseminated Greek philosophy, mobilized the
best craftsmen in the construction of palaces and mosques, and spread the use of
the compass, the manufacture of paper and advanced methods of irrigating crops.
The Arabs tended to be transmitters of techniques between the various regions
they controlled rather than innovators.   Islam was admired by contemporary
commentators  for its science, its orchards and its gardens, and  left prosperity
everywhere in its wake.

During that period, the Chinese were even more creative. They had invented
gunpowder, paper paste, the printing press four centuries before Gutenberg and

had constructed the first clocks. Their empire was ruled by absolute power; the civil service consisted of the well educated who were recruited by competition as objectively as possible. If one was to show an expert from Mars the technology of the 20th century and the state of different civilizations in the 10th century, he would no doubt say that China was the most likely place where the modern world would arise. And yet this was not so. Why?

In posing this question we confront the heart of the debate concerning futurology. Lacking a historical analysis that can explain the rise and fall of civilizations, futurologists are in an awkward position to evaluate the chances of our civilization in future centuries, and political leaders have to rely on dubious doctrines and irrational beliefs. The myriad of destructive ideologies which have occurred this century have led intellectuals to take refuge in skepticism, acting as if the human mind is completely incapable of any profound analysis of the major movements within society. This mistrust is understandable; but do not let me continue this despondency which in the end is incapable of producing anything of a positive nature. I will therefore attempt an analysis or suggest a line of research which can show the way.

A mere description of conflicts and oppressions is far from enough to understand the development of humanity. A description of history's failures does not make history, and if attempts to explain the past have so far failed, this is undoubtedly due to a lack of relevant information. We try to understand but look in the wrong places. The substance of a civilization consists of daily events and is formed out of what allows people to survive: techniques. Official history, the history of princes, their courts, and their battles -- precisely those events which have been made into a spectacle to be told -- needs to be avoided. Instead, we should look at what is happening backstage, even though little attention has been given to the development of techniques. This is a difficult, much fuller and less "visible" area, however determinant and gives form to social structures.

## 10.2   A STABLE CHINA: WHY?

Let's consider the example of China in the 10th century, mentioned above. The Middle Kingdom Empire was centralized and its inventions (gunpowder, paper paste, printing, clocks) came from within the Court and were used for the requirements and pleasures of the imperial rulers. Gunpowder was used in fireworks which the Chinese adored, and paper and printing for circulating imperial decrees and instructions. As for the clock, it suffered a sad fate which is worth narrating.

The first clock was built by the tantric Buddhist monk and mathematician,

Yixing, in 725. It was known as "the map of the spherical sky, in the shape of a bird's eye, powered by water", and was installed in the palace in full view of ministers and mandarins. In 730, candidates in the official examination for the selection of officials had to write an essay about the clock, but soon afterward the mechanism began to rust. Finally rendered unusable, it was removed to "the museum of the college of all sages."

The largest Chinese clock was the "cosmic machine of Su Song" constructed in 1092. It was ten meters high and topped by an armillary sphere from which one could observe the position of the stars. People said that the observations of the stars synchronized with the figures of the machine "like the two halves of a stone".

Reconstruction of Su Song's clock

From David S. Landes, Revolution in Time, Harvard Univ Press, 1983

**Fig. 10.1 Reconstruction of Su Song's Clock.**

The clock of Su Song worked from 1092 to 1126. When the Song dynasty was forced to abandon its capital, Kaifeng, the clock was dismantled and put together again in Peking where it continued functioning for a few more years. However, members of the political faction opposed to the Su Song's faction (he was a conservative) demanded the destruction of the clock for political reasons, because it symbolized the preceding era (the Yuan You era of two years earlier). The assistant director of the Imperial Museum managed to delay the decision by appealing directly to the prime minister, but when the new faction came to power soon afterward no one could prevent the destruction of the machine. A text of the period comments "what a shame!".

When technology is dependent upon the changes of mind and intrigues of those in power and is not rooted in the daily life of the ordinary population, it can be forgotten or destroyed. The Chinese only rediscovered this mechanical device for measuring time when the Jesuits brought over European clocks.

China was a rural civilization and agriculture and irrigation were the only technical areas in which knowledge was passed on to "civil society", as it is nowadays termed. The more eminent emperors were proud to have written a treatise on water management, as this area of knowledge put them in direct contact with the people. Other areas of knowledge were seen as trivial unless they concerned the maintenance of order needed to protect peasants from looting.

Actually, this turned out to be the weak link. Viewed over several millenniums, the history of China is a succession of oscillations between periods in which central power is strong and maintains justice and periods in which warlords, feudal outcasts, or highwaymen reign, pillaging the countryside and exacting ransoms. If power remained centralized in China, it is because although exploitative, it was a source of support. Imperial power was anchored in the Chinese mentality whereas Japan and Europe turned toward polycentric, feudal systems. China has forever existed in a balance of ebb and flow; the seasons, the harvests, power, Confucianism and Taoism, Yin & Yang -- each phenomenon and its opposite coexisted and became active in turn. The cycle of events could not be broken and therefore the very concept of progress had no place.

## 10.3 THE OPENING OUT AND TURNING INWARD OF ISLAM: WHY?

The case of Islam is also revealing. At its onset, this civilization was creative and lively, on the lookout for advances in science, technology and intellectual thought. For example, a Chinese captive taken in 751 at the battle of Talas taught the Arabs to manufacture paper. Papyrus was becoming rare and expensive but the need to write and communicate was deep-rooted. The first press was set up in

Baghdad in 795 and by the 10th century paper was becoming more prevalent than papyrus. At the time the population of Baghdad was over a million inhabitants while European cities such as Paris only reached 300,000 inhabitants in the 14th century. In the 12th century there were hundreds of paper mills functioning in the region of Fes in Morocco. Muslim society in the 10th century was already much more refined than European society and transmitted to Europe algebra, philosophy, medicine, irrigation, and navigational instruments. But then society became stilted and since the 12th century has taken delight in incantations, the absolute and repetition. Why?

Looking in more detail at the attitude of Arabs to technology, one gets the feeling that, despite their interest in some new inventions, they were not in agreement with the idea of overall progress. Many technical treatises are primarily compilations of older works, often from Roman times. In the field of agriculture, regional customs continued to flourish. Persians used their traditional form of plough while Egyptians used theirs. Advances in irrigation technique in the south of Spain were not disseminated to other regions. Local variations were paramount and Islam had just been superimposed over a mosaic of tribal societies with static cultures not open to compromise. Study and research were limited to a few centers of learning in Persia, Egypt and Cordoba, without permeating the population.

In the 12th century Cordoba was the location for one of the most beautiful love stories the human soul has ever known. Maïmonides - the Jewish sage and philosopher and author of the guide for lost souls, Averroès - a fountain of scientific, philosophical, and technical knowledge, and Alfonso X - the most tolerant of Christian kings, so tolerant that he was to be killed by his own subjects, were all living there at the same time. Ibn Arabi, the mystic luminary of Islam, author of the sublime phrase heralding modern science "Man is the eye of God", also dwelt there. They all preached the same message: "Knowledge (in its fullest sense) should be accessible to all men whatever their origin, their faith, their attachments, their richness or poverty". These wise men, from out of their diversity, and yet in agreement over the essential, conceived the early beginnings of human rights.

I believe this affinity between three such great minds was too powerful an experience for the time, a sort of spiritual earthquake. Out of fear, an isolationist reaction was unleashed as the three communities turned their back on love and went off in their own directions. The Spanish turned to reconquering their land, (the Inquisition), and then departed to plunder South America. Since then, their songs of love are platitudes or despairing litanies.

The Muslims, carried away by the rantings of Al Ghazali against the philosophers (falsafa) with their doubts and discussions, proclaimed the end of the "ijtihâd" (the path of perfection followed by believers to approach nearer to God) as a moratorium on study and research. They acted as if the world had come to a halt and could only continue existing in the remembrance of past glories. As for the Jews, they took up their wanderings again. Thus, while for Christians love was transformed into its opposite, destruction, Muslims projected it into the past and their society lost interest in the future.

Even now amongst Muslim traditionalists, important deeds carrying hopes for the future stay hidden. Revelation (i.e., creation) was perceived as close to blasphemy (bidda). This made it difficult (although not impossible), for innovation to flourish, because to innovate was to reveal that which was not yet visible.

The Koran is often used to sanction the maintenance of archaic tribal (sexist) customs dating from before the time of the prophet. Thus, while seeming to venerate Mohammed, in their actions his devotees deny that he transformed the world and ushered in a new era. According to the past critique of Ibn Khaldun, the decline of Islamic civilization is the concrete evidence of the irreality of its leaders, who prefer to chase their illusions rather than face up to reality. While people are suffering and becoming poorer, the rulers and their court are intoxicated by words and lavishness.

This behavior is not confined to the Islamic world. Indeed, it is seen so regularly in history that it could be called the "principle of irreality". This states that any ruling system, which becomes well-established and secure, tends to lose interest in the practical everyday reality of the life of its citizens and in intellectual and technological developments. It devotes the majority of its time to internal intrigues and keeps to its habitual ways of thought and explanation despite all evidence to the contrary, except when its survival or preservation of power is threatened. Thus, power and progress are rarely good bedfellows, as we will see below in the case of Europe.

## 10.4  12TH CENTURY EUROPE: THE FIRST LIFTOFF

Having looked at the stability of China and the decline of Islam, we can now consider why the extraordinary explosion of modern technological creation took place in Europe, a region which in the 10th century was inhabited by a rural population, governed over by an uncouth and energetic feudal class, and plunged into uncertainty by the collapse of the Empire of Charlemagne. Nothing seemed to suggest this was likely to happen.

While civilizations tend toward stabilizing their technological system and can stay in harmony for several centuries without changing their technology, Europe experienced a major destabilization at the end of the 11th century, the great agricultural revolution of the Middle Ages; another occurred in the 18th century, the industrial revolution; now it is entering another, this time world-wide: the immaterial revolution.

I believe the weak point which allowed change to enter this civilization while others resisted was the absence of power. Why, for example, did knights go off on the Crusades? Not just for the official reasons given in the writings of the clergy of the era, but also for very practical reasons related to the conditions of the time. From the 11th century onward, disagreement was becoming apparent between the two components of the ruling class: temporal feudal power on the one hand, and the spiritual power of the church and the monasteries on the other hand. The numerous idle offspring of the knighthood engaged in pillage, ruining crops by riotous gallops through the fields and even ransacking monasteries. After various unsuccessful attempts to control these excesses, the Church came up with the idea of the Crusades. What a brilliant idea to channel the excess energy and the thirst for ideals of these predatory young men. Initiatives began to flourish in the fortuitous absence of this feudal class. The administrators of rural holdings started to go to market (it had been forbidden), saved money, invested it, brought new land into cultivation and tried out new crops. Freed from its ruling class, Europe began to see a new entrepreneurial spirit emerge.

Concomitantly, the monasteries found themselves in financial difficulties. They had gathered a sizable population (some 25,000 monks in the Order of Cluny alone) and their lands were vast. But singing hymns and praying daily for seven hours while refusing to cultivate the land themselves, the clergy had confided its management to peasants. However, the peasants devoted their energies to hiding crops and even traces of ownership. So, Cluny, for instance, without sufficient resources and burdened by excessive expenses, was no longer able to pay what it owed.

Financially vulnerable, the Church's spiritual hegemony was at the same time also under challenge. Brought by merchants, a heresy from the East spread through Northern Europe before gaining the support of the Count of Toulouse and the "Albigensians". The Inquisition was set up to fight against heresy. Heirs to a long-standing dualist tradition even earlier than Christianity,[1] the Cathar heretics, claimed that one did not need the Church to be close to God and even more they suspected Rome to be a manifestation of evil, especially as it claimed to represent a poor God while itself hoarding immense riches. Indeed the bishops, abbots, and monks of this period lived in great style, light-heartedly spending the income from their domains and engaging in varied escapades with little danger of being called to justice as their privileged and sacred status placed them above the law. Just as dangerous for the clerical institution was the growing internal controversy initiated by Abélard.  This lively, spirited and turbulent monk took part in oratorical disputations in front of his students at the Montagne Sainte Geneviève in Paris.  He discussed sacred texts without always referring to the Church commentaries on them.  While this can seem of little importance to us, it was a crucial issue at the time, as the right to comment freely meant no less than the advent of freedom of thought.  Abélard's "disputations" were the foundations of the modern concept of a university, developed from Averroes and the sages of Cordoba.  The flame of philosophical doubt had caught light in Europe and would never again be extinguished.

The Church felt the need to turn to an ascete employing an iron fist.  Bernard of Clairvaux, the future St. Bernard, imposed his ideas from 1117 onward.  The Cistercian revolution meant working with one's hands as the original rules laid down by St. Benoît had intended, fleeing the towns, the new Babylons, dens of corruption, getting rid of luxury and ornaments and helping the rural population practically.  Knowledge accumulated in the manuscripts of the cluster of monastic orders, the sole medium at the time for the circulation of learning, was mobilized to save Cluny and ensured eventual success.  Hundreds of monasteries rallied to the new doctrine and new establishments were constructed in virgin territory, or as was stated metaphorically, "in the desert".  A total of 700 sister abbeys were built in two centuries, with one a week during the highest period of growth (1145-50).  They disseminated technical knowledge into the surrounding rural areas.  The selection of seeds, breeding of animals, and spread of mills, not only as a source of energy for grinding but also for sawing wood, pressing cloth and activating the bellows of the forge date from this period.  So do the iron plough and the horse's yoke necessary for wide-scale land clearing and cultivation.  Markets developed and became international.

In the 13th century, prefiguring the advent of capitalism, international trade developed around the Baltic towns of Lübeck, Brême, Köln, Danzig, Goslar, Hamburg, Lunebourg, Reval, Riga, Rostock, Stralsund.  These towns formed the Hanseatic League which gave rise to a form of isonomic government and laid down strict rules for commercial activities.  Improved navigational techniques were employed.  The "cogges" could carry up to 120 tons and were equipped with the first stern post rudders.  A wreck found in the port of Bremen in 1962 was 23.5 meters long, 7.5 meters wide with a waterline of 2 meters.  The crew would have been 15-20 men.  It sank just a short time after its completion in 1380.

These boats were the forerunners of the vessels which were to set out to conquer the Americas in the succeeding centuries.  The affection of German cities for their fairs dates from this period.  Prosperity burst out everywhere and the population doubled between 1100 and 1300.

## 10.5 DECLINE, RENAISSANCE, AND INDUSTRIAL REVOLUTION

Unfortunately, this growth ended disastrously. By the beginning of the 14th century population density had reached 40 inhabitants per square kilometer, the maximum sustainable by this rural technology. Climatic variations were enough to cause an initial famine (1316) and the great plague of 1348 decimated an already weakened population. In one year it killed one-third of the population of Europe. It kept recurring and was endemic until about 1475 at which stage the One Hundred Years War commenced. In all, two centuries of misfortune marked the European conscience. It seemed like some original, mysterious sin, a breach between people and the natural order which it was important to close.

This collapse was accompanied by a hardening of society. Technology was no longer freely available, guilds re-emerged and areas of professional domination were delineated. Innovation became less and less possible as links in the chain of prohibitions were reinforced. The means of production were expropriated by existing institutions which nonetheless remained a recourse against misfortune, for lack of any better alternative.

The population diminished by half between 1300 and 1500 returning to levels like those prior to the period of great medieval prosperity. What we term the Renaissance in fact marks the end of this great, painful decline. The fundamentals had already been invented. Great engineers such as Leonardo da Vinci gave concrete form to designs already known. The basics of techniques did not change until the 18th century, except in two respects:

(1) *Expansion.* The world enlarged with the conquest of America and especially with the establishment of the first global trade routes. This represented the application world-wide of the system originally conceived by the Hanseatic League in the Baltic.

(2) *Communication.* Printing's first effect was on religion because, despite the Inquisition, the Church could no longer prevent believers from reading and thinking about the sacred texts for themselves. The Church was unable to control the spread of Protestantism which occurred as a result of the diffusion of the holy texts. Two centuries later printing had repercussions on technology. Through the publication of the Encyclopaedia (24,000 copies) knowledge that before was jealously guarded by the guilds became available to the public and stimulated the extraordinary creativity of the Industrial Revolution.

This revolution in the 18th century had disconcerting similarities with the one that occurred during the Middle Ages. From the time of Louis XIV onward, the ruling class was weak and divided. While still a youth, this French monarch countered the *Fronde* by attracting the nobility to his court, a sumptuous diversion, and a wonderful decoy. In so doing, he separated them from their lands which they were meant to administer and thus their stewards took advantage. By the time the nobility had spent two generations in the diversions of the court, they had become incompetent -- and the clergy were little better.

Within this ruling class, cut off from reality, living in a surreal world, emerged an innovative minority calling for a return to fundamentals, just as the Cistercians of Saint Bernard had previously done. This was the movement of the philosophers whose ideas inspired the French Revolution. Here again, the destructuration of power preceded technical innovation and its offshoot, economic prosperity.

In England where the industrial revolution preceded that of France, one can see the same weakening of central power exacerbated by economic crisis. The competition of the Indian silk trade -- already using cheap labor -- threatened the British wool trade. The entire economic chain from the sheep to rural weaving was endangered. Prohibitions and regulations were laid down and merchant cargoes were burnt. But the pressure remained, resulting in a restructuring of agricultural land holdings by the rising landlord class. This opened the way for industrial inventions, commencing initially in the textile industry with mechanical spinning and weaving. So the competition from India was defeated by technical advances in machinery.

## 10.6 THE MIDDLE OF THE 19th CENTURY

In the Age of Enlightenment, as in the 12th century, the ruling class became energized by a rescue plan. In the Middle Ages it involved channelling violence and rediscovering the true values of survival: a love of nature and the art of living in harmony with it. In the 18th century the largely British-inspired movement of the philosophers also had its grand plan: to mobilize science and its progeny, technology, to help humanity escape from the misery to which it had seemed tied for so many centuries. In both cases civilization stirs itself and technology makes a shift, but the people do not follow this movement.

Several decades passed and the results were not those expected by the initiators. In the first half of the 19th century industry grew, and the countryside emptied to the advantage of the outskirts of large towns. Some individuals became rich but the majority had to face up to developments which made them even poorer and more dependent on others.

In the 1848 revolts in Europe, the ruling class lost its illusions. Faced with a burgeoning lower class living in appalling poverty and extremely unhealthy conditions, two currents of opinions emerged; one humanist and the other conservative. The humanists said: "We can't let human beings live in such awful conditions, it's intolerable". They were right. And the conservatives added: "Watch out, they are becoming dangerous!" They were also right. Both arrived at the same conclusion: something had to be done.

Tremendous means were mobilized. From a frightened, narrow-minded bourgeoisie that was preoccupied with hoarding its money, litigating endless disputes, and sending debtors to prison, emerged a modern and powerful bourgeoisie looking to the future, investing massively in building railways, department stores, banks, heavy industry, roads, the Suez and Panama canals. They took enormous risks, demonstrated a world-wide vision and to fulfill their plans, created a pool of liquidity through the "transformation of credit".

Our ancestors structured towns and structured minds. Great Britain under Queen Victoria, Germany during Bismarck's time, and France under Napoleon III saw the same changes: investments in urban development, education and social control on a tremendous scale. And it was successful! Instead of the revolution foreseen by Marx, Europe in 1900 lit up the world, and despite two world wars Western Europe in the 20th century prospered as never before in its history.

The major European cities reached the 1900s with thoroughfares wide enough for cars (although laid out well before their invention), and a population educated to a standard sufficient for its role in the industrial economy. Europe's present development is the result of the urban structuring of the last century which, incidentally, was envisaged with an eye on maintaining social order, and the educational structuring of that century which was not without thoughts of social control. Popular technical culture, out of synchronization at the onset of the industrial revolution, is once again in phase.

The educational material of the period 1850-1900 reveals that the aim was not solely to teach reading, writing and counting. People were taught by projections of "magic lanterns" that they had to have a family, a home, send their children to school, be employed, give up drinking and conform to the middle class morality of the period. They were also taught the most practical and everyday results of science and technology via "practical lessons".

This reference to the past suggests a future rupture in society. If similar causes produce similar effects then when the situation becomes intolerable, the ruling class will again be seized with fright and take the actions necessary to remedy the situation. The reaction of the last century presents a universal characteristic - by structuring space, they structured minds. It was expressed in vast programs and carried out with the most advanced technologies of the time. We can conjecture that the response to the increase in danger at the beginning of the next century will be similar, although relying on much more powerful techniques. History brings these visions of catastrophe into perspective; humanity has stood up to worse and survived.

**Fig. 10.2  In space, gigantic fields of solar receptors gather energy which is then transmitted to earth by microwave.**

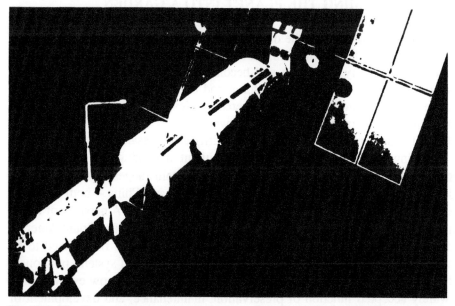

**Fig. 10.3  The first permanent space stations are directly supplied with solar energy.**

Fig. 10.4 Guided by men and women, the mining operations on Mars are carried out by robots.

## 10.7 TRANSFORMING THE PLANET INTO A GARDEN

The great richness of life and the diversity of our ecological heritage needs to be preserved. Further, humankind must see that as it can now control nature, it also assumes responsibility for nature and becomes the guardian of life. It has the power to destroy life but also to preserve and enrich it. A world program for nature reserves, the protection of endangered species, conservation of our genetic heritage, and reforestation needs to be drawn up.

Global management of water, including irrigation, desalination, purification, and recycling, needs to be implemented without damaging the environment. Dams need to be installed in the two largest mountain ranges in the world, the Andes and especially the Himalayas adjacent to India and China -- both countries potentially large consumers of electricity in the next century. A network of lakes, dams, and canals in Africa, the Indian peninsula (Bangladesh), and South America need to be constructed, and the courses of the great Siberian rivers need to be sensibly adapted to counter global warming. The agricultural system, previously exclusively concerned with food production for the market, is now evolving into the widely accepted function of preservation, maintenance, and development of nature. The exploiter becomes transformed into an artist chaperoning the earth's

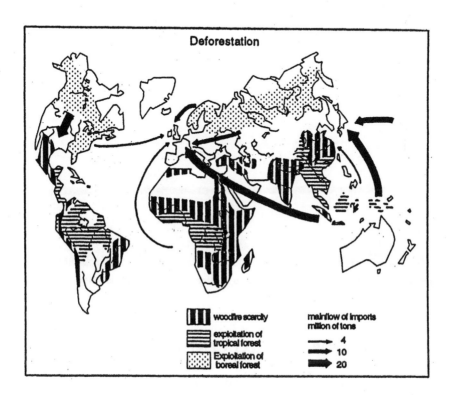

**Fig. 10.5  Deforestation around the world.**

fertility.  The reconquest of areas desertified by humankind through over-
exploitation, deforestation, or abandonment represents one of the greatest
challenges of this development.

   The institutional framework for these tasks will not only consist of state
administrations but also of an appropriate world legislation commanding states
and a transnational network of agencies (modelled on the water-regulatory
agencies) led by professionals, levying taxes on damages to nature, providing
public services and using their resources in the interest of nature.

**Fig. 10.6  In Biosphere 2, humans endeavor to master a complete and isolated ecosystem.**

## 10.8  PROGRAMS FOR THE NEXT CENTURY

Viewed on the scale of centuries, the settling of a new technical system generates social trouble by slowly excluding the previous working force.  Our cognitive revolution, which started during the 1980s and will expand worldwide during the next century, does not escape this rule.  The ruling classes are not yet worried enough to change their strategy.  They are certainly aware that social exclusion and urban insecurity are growing, but still believe that they can respond defensively to them.  They will not be prevailed upon to implement constructive projects until the law and order option has been exhausted.  Our estimation is that the transition to a society of education and major projects will occur between 2010 and 2020.

Already the need for economic revival via a program of civil, peaceful, and generous initiatives is becoming more apparent all the time.  Instead of letting millions of unemployed without hope continue to build up, we need to make work satisfying again and find a way to use everyone's talents in constructing a new and better world for our children.  Voices are already being raised to propose audacious schemes comparable to those of the 1960s when we dared proclaim "In ten years, we'll have landed on the moon". But technological projects will not be enough. The world needs more humanity.

In proposing these programs, aren't we in contradiction with innovation, which develops out of micro-initiatives? Or with the flourishing and estimable school of thought which defends the "small is beautiful" thesis, corroborated by the decentralization scenario which we have described as a consequence of the communications network? After a period in which megalomaniac engineers reckoned only in millions of dollars and constructed "industrial cathedrals in the desert", two out of three of which no longer work, we then experienced the opposite, the cult of austerity and autonomy, as expressed in the pedal wheels of the disciples of Gandhi and the goatskin windmills developed for the Third World by militant European ecologists. There are fashions in technology as in everything else. They are only a testimony to the indecisiveness of our judgement. The idea of technology itself cannot be bound by our prejudices. It concerns the essence of existence itself, which cannot be controlled by humankind.

In my opinion there is no contradiction. The choice is not between small and large. The idea of programmation is fractal and expresses a common attitude, whatever the size of the project. The largest deeds come from small groups. This will be even more true in the future as there will be less need for mobilizing a vast army of participants. Repetitive and mechanical tasks will be taken over by machinery.

The idea is not to propose programs that take refuge in technology, with the intention of blotting out and forgetting the real difficulties which people face in their daily lives. Modern means of measurement, forecasting, and communication allow schemes to be evaluated and social dialogue to occur where previously one used force, secrecy, or surprise to get one's way. The risks of technological frenzy are shown in the ecological damage caused by certain large projects (such as the Aral Sea and Aswan dam). But is this a reason to give up all projects? Certainly not! Each period had its own, the pyramids of antiquity, the cathedrals of the Middle Ages, the palaces of royalty. They express themselves at all levels from the individual to the species as a whole.

Past experience shows that finance is made available on an all-or-nothing basis. Either money is brusquely refused or the floodgates are opened without any restraint so as to provide without delay huge amounts of money for projects that have suddenly become urgent. In the latter case, if the groundwork has not been prepared, politicians are liable to be swayed by illusions and finance poorly planned projects just to give the appearance of doing something.

There is a logic to infrastructures that is neither that of the market nor that of prestigious projects. This is the logic of public utility which consists of providing the necessary public services in appropriate conditions of availability at optimum cost to the community. This logic applies at a municipal level for public transport and other local services; at a global level for the protection of whales, lemurs, bears, parrots, and other endangered species; for satellite weather forecasts; and the standardization of industrial products.

Currently, people of goodwill have a few years ahead of them to prepare programs for the next century. It is important to start work now as the design of quality infrastructures takes time both in terms of engineering and of negotiations. Computerization allows more rapid progress to be made and more research into variants. Once stored in memory, a project becomes a modifiable virtual universe through which one can wander in images. It becomes presentable to future users prior to actually being built. Discussions can then take place and a common consensus can be found. If we set aside the time, we do have the means to carefully study and discuss our programs in all their ramifications and to integrate them in a more social and democratic process. The imaginary should precede the real and not follow it. We need to start work now on the infrastructures of the next century.


## 10.9  CONCLUSION

The 21st century will be the century in which I believe the unity of the human species will occur. Due to change in technical systems, the future can no longer be thought as a continuation of the past. Our era presents two radically new modes of operation never before experienced by humankind in its two million years of existence.

The first mode is the ability to instantaneously communicate from one side of the planet to the other. By the first quarter of the next century, communication will become audio-visual and will reach even the most disadvantaged groups of people. Almost all countries will cross the threshold of ten telephone lines per one hundred inhabitants before 2020. This web of communication means it will no longer be possible to control civil society. The very concept of power will have to be reconsidered.

Following industrial society, we will enter into the cognitive civilization. Now, to each technical state specific forms of organization correspond: for the hunter-gatherers, the tribe; for the farmers, the village and feudalism; for industry, the large company and the nation-state. For this new technical system, the normal form of socialization is the small firm, and for community requirements, associations, foundations, local authorities and professional organizations. All these organizations are constructed and organized in line with cognitive needs.

They must balance their accounts for fear of insolvency. They are the emanation of the human conscience, translated into volition. Each generation can create them, animate, them, suppress them, and reorganize them anew according to its own desires.

The second mode of operation results from the perception of the limits of the world and the fragility of life. The last remaining wild forests are threatened. The perspective of their disappearance is a source of distress. After a period of unrestrained consumption, humans start to feel responsible and seek to limit energy waste, pollution, to preserve flora and fauna, and to heal nature's wounds.

Humanity is increasingly exercising birth control. Demography will be stable in a little less than one hundred years at around thirteen billion inhabitants, that is, slightly more than double the present population. It will be the century of the Feminine. The values of preservation of life, harmony, and balance will become more important than the previous masculine values of conquest and authority, which fitted in with the periods of unbridled expansion of the species.

While men and women become definitively responsible for nature, as its "gardener" they also transform. It becomes a "techno-nature". From now on, all the environment is re-created. it is imagination given form. The issue of technology now presents itself in other terms. It is no longer about just providing useful answers to specific needs but about re-creating conditions favorable to the spread of life in all its forms.

These two characteristics of our time -- instantaneous communications and fragility of life -- give birth to a planetary consciousness. At the same time, they call forth a continual re-creation of the world, focusing on innovation as the central question in controlling this poorly defined future. However, the culmination of this scenario in the next century, creative freedom, will occur only after a long march riddled with trials and tribulations.

### Notes

1.  The Zoroastrians were already dualist (800 B.C.). For them the actual world was the setting for a struggle between the forces of light and darkness. The confrontation of these two "supernatural" principles would finish with the triumph of light, but only at the end of time.

# INDEX

## FINITE ELEMENT PROGRAMS IN STRUCTURAL ENGINEERING AND CONTINUUM MECHANICS

CARL T. F. ROSS, Professor of Structural Dynamics, Department of Mechanical and Manufacturing Engineering, University of Portsmouth

**ISBN 1-898563-28-4**            **650 pages**                    **1996**

Covers finite element programming in a wide range of problems in mechanical, civil, aeronautical and electrical engineering. Comprehensive, it ranges from the static analysis of two- and three- dimensional structures to stress analysis of thick slabs on elastic foundations, and from two- and three-dimensional vibration analysis problems to two-dimensional field problems, including heat transfer and acoustic vibrations.

The 24 printouts of powerful and valuable engineering computer programs written in QUICK BASIC analyse engineering design and manufacturing problems by the finite element method.

**"Computer programs for finite element analysis .... .students and lecturers may find them of value"**

*The Structural Engineer* (Professor I.A. Macleod, Strathclyde University, Glasgow)

## FINITE ELEMENT TECHNIQUES IN STRUCTURAL MECHANICS

CARL T. F. ROSS, Professor of Structural Dynamics, Department of Mechanical and Manufacturing Engineering, University of Portsmouth

**ISBN: 1-898563-25-X**            **224 pages**                    **1997**

This undergraduate and postgraduate text will serve for courses in mechanical, civil, structural and aeronautical engineering; and naval architecture. It is written in a step-by-step methodological approach, so that undergraduate readers can acquire knowledge, either through formal engineering courses or by self-study.

**"All Carl Ross' previous books are very clear and well written. This thoroughly interesting text is no exception. The worked examples and practice problems are particularly useful. I will continue to recommend Professor Ross' books to my students. For anyone requiring an introduction to finite element analysis, this text is excellent"**

*Journal of Strain Analysis* (Dr. S.J. Hardy, University College of Wales, Swansea)

**"The finite element method emphasizes practical aspects of the method, deliberately written in nonrigorous fashion ... recommended for upper-division undergraduates, graduate students, and practicing engineers".**

*Choice* (USA), (D.A. Pape, Alfred University, USA)

## DYNAMICS OF MECHANICAL SYSTEMS

CARL T. F. ROSS, Professor of Structural Dynamics, Department of Mechanical and Manufacturing Engineering, University of Portsmouth

**ISBN: 1-898563-34-9**                    **306 pages**                    **1997**

A fundamental introduction for undergraduates reading mechanical, civil, structural and aeronautical engineering, and naval architecture. The step-by-step and methodical approach is aimed to help students who find difficulty with mathematics and Newtonian Physics.

Rectilinear and curvilinear motion are considered, with emphasis on Newton's Laws of Motion. Special features are the introduction and painstaking use of the Stroud system of units, and a chapter on gyroscopes.

Carefully explained worked examples are set out in detail, with accompanying diagrams. All chapters have sections on "problems" and problem/exercises (with solutions and hints) test comprehension, or assist self-study.

*Contents*: Introduction to Statics and Dynamics; Kinematics of Particles; Kinetics of Particles: Force, Mass, Acceleration; Kinetics of Particles: Work, Energy, Power; Kinetics of Particles: Momentum and Impulse; Kinematics of Rigid Bodies; Kinetics of Rigid Bodies; Gyroscopic Theory and Applications; Free and Forced Vibrations; Appendices: Vector Algebra, and Mass Moments of Inertia.


## TEACHING AND LEARNING MATHEMATICAL MODELLING

Editors: S.K. HOUSTON, University of Ulster, Northern Ireland; W. BLUM, University of Kassel, Germany; IAN HUNTLEY, University of Bristol; N.T. NEILL, University of Ulster, Northern Ireland

**ISBN: 1-898563-29-2**                    **402 pages**                    **1997**

Mathematicians from ten countries contribute interdisciplinary applications in mechanics and engineering, computing science, traffic control, business studies, and mathematics (fractals and analysis). Tertiary and secondary levels are covered.

Philosophically and creatively they discuss innovation and assessment, and teaching and study, at all levels. The interdisciplinary nature of the topics reflects their use in such varied areas of application as mechanics and engineering, patient flow through hospitals, computing science, traffic control, business studies, and mathematics (fractals and analysis), all pointing to a wide choice of future careers.

Printed and bound by CPI Group (UK) Ltd, Croydon, CR0 4YY